T0201259

Solving Partial Differential Equation Applications with PDE2D

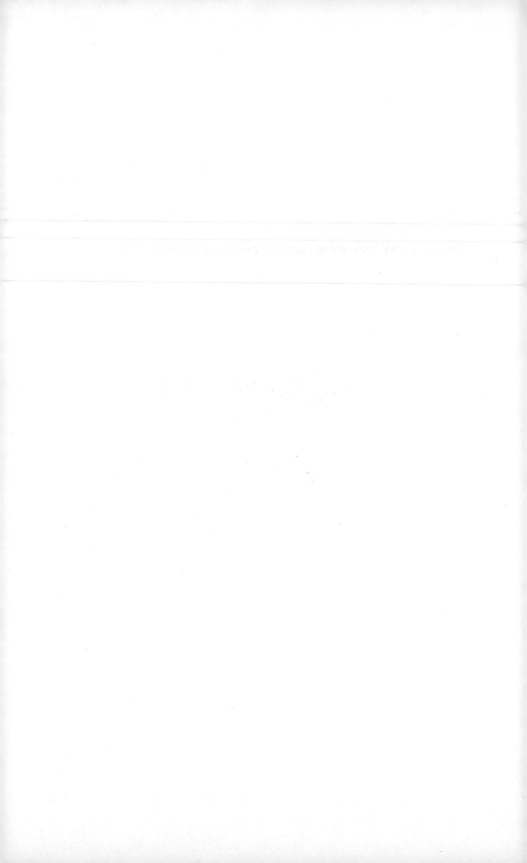

Solving Partial Differential Equation Applications with PDE2D

Granville Sewell

Mathematics Department
University of Texas El Paso
El Paso, TX, USA

Registered Office
John Wiley & Sons, Inc., 111 River Street, Hoboken, NJ 07030, USA

Editorial Office
111 River Street, Hoboken, NJ 07030, USA

For details of our global editorial offices, customer services, and more information about Wiley products visit us at www.wiley.com.

Wiley also publishes its books in a variety of electronic formats and by print-on-demand. Some content that appears in standard print versions of this book may not be available in other formats.

Library of Congress Cataloging-in-Publication Data

Names: Sewell, Granville, author.
Title: Solving partial differential equation applications with PDE2D /
   Granville Sewell.
Description: Hoboken, NJ : John Wiley & Sons, 2018. | Includes
   bibliographical references and index. |
Identifiers: LCCN 2018018589 (print) | LCCN 2018031767 (ebook) | ISBN
   9781119507956 (Adobe PDF) | ISBN 9781119507963 (ePub) | ISBN 9781119507932
   (hardcover)
Subjects: LCSH: Differential equations, Partial–Numerical
   solutions–Computer programs.
Classification: LCC QA377 (ebook) | LCC QA377 .S4643 2018 (print) | DDC
   515/.353028553–dc23
LC record available at https://lccn.loc.gov/2018018589

Cover design by Wiley
Cover image: © Courtesy of Granville Sewell

Set in 10/12pt Warnock by SPi Global, Pondicherry, India

Printed in the United States of America

V10004135_082918

Contents

Preface

This book looks at a wide range of ordinary and partial differential equation (PDE) applications. Students using this text will actually solve many interesting science and engineering applications using PDE2D, an easy-to-use, general-purpose PDE solver developed by the author over a 40-year period, which is free with the purchase of this book. They will learn to derive and solve the ordinary or partial differential equations, with boundary and initial conditions, for many time-dependent, steady-state, and eigenvalue applications, including diffusion, heat conduction and convection, image processing, math finance, fluid flow, elasticity, and quantum mechanics, in one, two, and three space dimensions. That PDE2D can be used to solve a wide variety of applications is evidenced by the list of over 250 journal articles and books at www.pde2d.com, in which it has been used to generate some or all of the numerical results.

Much of the material in the book was developed for two graduate courses, "Seminar in Applied Mathematics" at Texas A&M University and "Advanced Scientific Computing" at the University of Texas El Paso, but the book could also be used as a supplementary text for a number of science or engineering courses in which PDE applications are studied. It could also be used as a reference by individual students or researchers interested in using PDE2D to solve their specific applications.

Some documentation on the mathematical algorithms used by PDE2D can be found in Appendix B. The focus in this book, however, is on use of PDE2D, not the mathematics behind it, and students are not primarily learning about numerical methods, though they will learn some things, but rather about modeling real-world applications using differential equations.

The book starts with some simple ordinary differential equation problems in Chapter 1, which give the student a chance to become familiar with PDE2D before proceeding to more difficult problems, and ends with the solution in Chapter 9 of a very difficult nonlinear problem, which requires a moving adaptive grid because the solution has sharp, moving peaks.

The Windows/GFortran version of PDE2D is available at no cost with the book purchase: go to the "Free with Book" page at www.pde2d.com.

Linux and Mac OS X versions are available for a fee, also from www.pde2d.com. All three versions require the GNU GFortran compiler, but this can be downloaded from http://gcc.gnu.org/wiki/GFortranBinaries for free. Instructors using this book as a required text for a university class, and their students, can also obtain the Linux or Mac OS X version at no cost.

Each problem comes with some graphical or numerical output so the student can tell when his/her program is working correctly. The example programs can be downloaded from www.pde2d.com.

I

Introduction to PDE2D

I.1   The Collocation and Galerkin Finite Element Methods

PDE2D is a general-purpose partial differential equation (PDE) solver that solves very general systems of nonlinear, steady-state, time-dependent, and eigenvalue PDEs in 1D intervals, general 2D regions (see Figure I.1), and a wide range of simple 3D regions (see Figure I.2), with general boundary conditions.

PDE2D uses a collocation finite element method for 3D problems, and either a collocation or Galerkin finite element method can be used for 1D and 2D problems. This book is primarily concerned with using PDE2D to solve applications, so we will not attempt in this section to explain the mathematics behind either method. Readers who want to know more about the mathematics should read Appendix B and for more details, *The Numerical Solution of Ordinary and Partial Differential Equations* (*Third Edition*) (Sewell 2015). Here we are only concerned with helping users decide which method to use.

There are two major differences between the collocation and Galerkin methods, as far as users are concerned:

1) The Galerkin method requires that PDEs be in "divergence" form. For 2D steady-state problems, this means it must be possible to put them into the form:

$$\frac{\partial}{\partial x} A_1(x, y, U1, U1_x, U1_y, ..., UN, UN_x, UN_y)$$

$$+\frac{\partial}{\partial y} B_1(x, y, U1, U1_x, U1_y, ..., UN, UN_x, UN_y) =$$

$$F_1(x, y, U1, U1_x, U1_y, ..., UN, UN_x, UN_y)$$

$$. \quad =$$

$$. \quad =$$

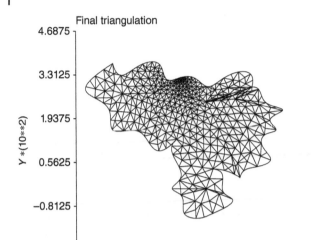

Figure I.1 Triangulation of Venezuela.

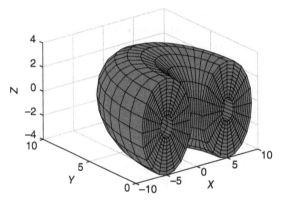

Figure I.2 3D grid, hollowed-out half torus.

$$\frac{\partial}{\partial x} A_N(x, y, U1, U1_x, U1_y, ..., UN, UN_x, UN_y)$$

$$+\frac{\partial}{\partial y} B_N(x, y, U1, U1_x, U1_y, ..., UN, UN_x, UN_y) =$$

$$F_N(x, y, U1, U1_x, U1_y, ..., UN, UN_x, UN_y)$$

where $U1(x, y), ..., UN(x, y)$ are the N unknown functions.

The boundary conditions are either "fixed",

$$U1 = FB_1(x, y)$$
$$. \quad = \quad .$$
$$. \quad = \quad .$$
$$UN = FB_N(x, y)$$

or more general "free" conditions,

$$A_1 N_x + B_1 N_y = GB_1(x, y, U1, U1_x, U1_y, ..., UN, UN_x, UN_y)$$
$$. \quad = \quad .$$
$$. \quad = \quad .$$
$$A_N N_x + B_N N_y = GB_N(x, y, U1, U1_x, U1_y, ..., UN, UN_x, UN_y)$$

where (N_x, N_y) is the unit outward normal vector on the boundary. The equations for time-dependent and eigenvalue problems, and 1D problems (listed in Appendix C), must be put into a similar divergence form.

New users sometimes find the Galerkin "free" boundary condition format confusing at first, especially when there are fixed and free boundary conditions on the same boundary, such as in Problem 2 of Chapter 2.

The collocation method allows a simpler, apparently more natural form, which for 2D steady-state problems means

$$F_1(x, y, U1, U1_x, U1_y, U1_{xx}, U1_{yy}, U1_{xy}, U2, ...) = 0$$
$$. \quad = .$$
$$. \quad = .$$
$$F_N(x, y, U1, U1_x, U1_y, U1_{xx}, U1_{yy}, U1_{xy}, U2, ...) = 0$$

The boundary conditions also seem to be more natural:

$$G_1(x, y, U1, U1_x, U1_y, ..., UN, UN_x, UN_y) = 0$$
$$. \quad = .$$
$$. \quad = .$$
$$G_N(x, y, U1, U1_x, U1_y, ..., UN, UN_x, UN_y) = 0$$

Periodic and "no" boundary conditions are also permitted.

The equations for time-dependent and eigenvalue problems, and 1D and 3D problems (listed in Appendix C), have similar (nondivergence) formats.

For problems such as the Black–Scholes equation (4.9), the collocation format is simpler. It requires some additional work to put the Black–Scholes equation into divergence form, and similar, but nonlinear, math finance equations are sometimes difficult or occasionally even impossible to put into divergence form, so the collocation method is preferred for problems such as these.

However, many applications, such as the elasticity equations (5.2) and the minimal surface equation (8.1), are already in divergence form, and it may require additional work to put them into a form suitable for the collocation method (see Problem 1 of Chapter 8). In fact, even the apparently unnatural Galerkin free boundary condition format is actually very natural for many applications, including elasticity problems with boundary forces given (5.5), which are much simpler to handle with the Galerkin finite element method. Furthermore, some problems, such as diffusion in a composite material (3.5), where the diffusion coefficient is discontinuous at an interface between materials, can be easily handled using the Galerkin method (see also Problem 6 of Chapter 1), while the collocation method cannot handle such composite problems at all.

2) The other major difference between the PDE2D Galerkin and collocation finite element methods, as far as users are concerned, is in the regions they can handle and how they are defined by the user. Since PDE2D only offers a collocation option for 3D problems, and for 1D problems the "regions" are just intervals (which both methods handle easily!), this difference is really only important for 2D problems.

For 3D problems, the PDE2D collocation method can only handle regions that can be parameterized (smoothly) as $X = X(p1, p2, p3), Y = Y(p1, p2, p3), Z = Z(p1, p2, p3)$, with constant limits on the parameters $p1, p2, p3$ (similarly for 2D problems). This means it can handle "a wide range of simple regions," but certainly not complicated regions such as Venezuela. The torus section of Figure I.2, used in Problem 4 of Chapter 3, was parameterized using the option ITRANS=−3 with equations

$$X = (5 + p3 * cos(p2)) * cos(p1)$$
$$Y = (5 + p3 * cos(p2)) * sin(p1)$$
$$Z = p3 * sin(p2) \tag{I.1}$$

where $p1$ (the toroidal angle) varies from 0 to π (only half of a torus is generated), $p2$ (the poloidal angle) varies from 0 to 2π, and $p3$ (the radial distance from centerline) varies from 1 to 4. Although regions are defined in terms of parameters $p1, p2, p3$, the PDEs, boundary conditions, and everything else are still defined using rectangular coordinates. For example, Laplace's equation can be written $Uxx + Uyy + Uzz = 0$ no matter how the region is parameterized.

If the collocation method is used to solve the minimal surface Problem 1 of Chapter 8, the following parameterization could be used, where $-1 \leq p1 \leq 1, -1 \leq p2 \leq 1$):

$$X = p1$$
$$Y = p2 * (1 + p1^2) \tag{I.2}$$

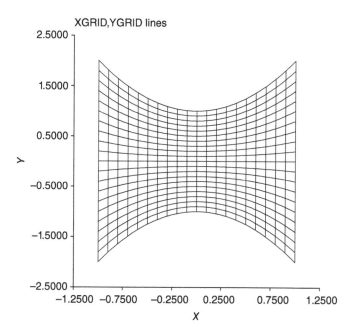

Figure I.3 Collocation grid for Problem 1, Chapter 8.

Figure I.3 shows the collocation grid generated for this problem, with $NP1GRID = NP2GRID = 21$.

The PDE2D Galerkin method, on the other hand, can solve problems in completely general 2D regions, with arbitrary curved boundaries (see Figure I.1). However, the user must (if the INTRI=3 option is chosen) supply an initial triangulation of the region, which normally consists of the minimal number of triangles needed to define the region. Inputting the initial triangulation is sometimes painstaking, but once the initial triangulation is defined, it can be automatically refined and graded by PDE2D. An initial triangulation of the minimal surface region of Figure I.3 is shown in Figure I.4.

However, if the region is rectangular or simple enough to be defined using parametric equations with constant limits on the parameters p, q (i.e. if it can be handled by the collocation method![1]), PDE2D can generate the initial triangulation automatically. Figure I.5 shows the initial triangulation generated by PDE2D if the option INTRI=2 is chosen and the parameterization (I.2) is used (with $p1, p2$ replaced by p, q). Notice that each grid square is divided into 4 triangles. For some regions, this will save the user a large amount of work.

---

1 Except that the parametric equations that define the region for the collocation method must be continuously differentiable, while they only need to be continuous for the Galerkin method.

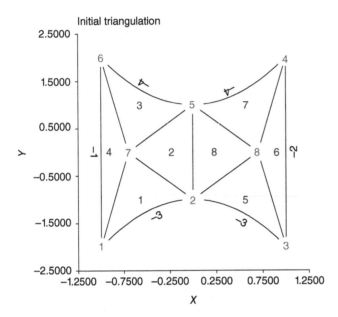

Figure I.4 Manually supplied initial triangulation (Galerkin method).

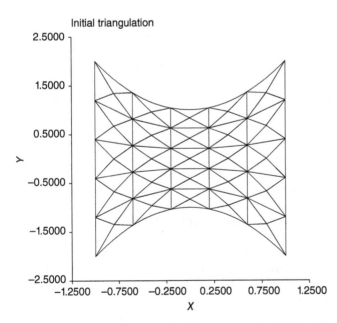

Figure I.5 Automatically generated initial triangulation (Galerkin method).

I.2 The PDE2D User Interfaces

There are two user interfaces, a graphical user interface (GUI) (Figure I.6, invoked by "pde2d_gui *name*"), which can be used to access the collocation finite element methods, and an interactive driver (Figure I.7, "pde2d *name*"), which can be used to access both the collocation and Galerkin finite element algorithms. The GUI is extremely easy to use, chooses more defaults, and

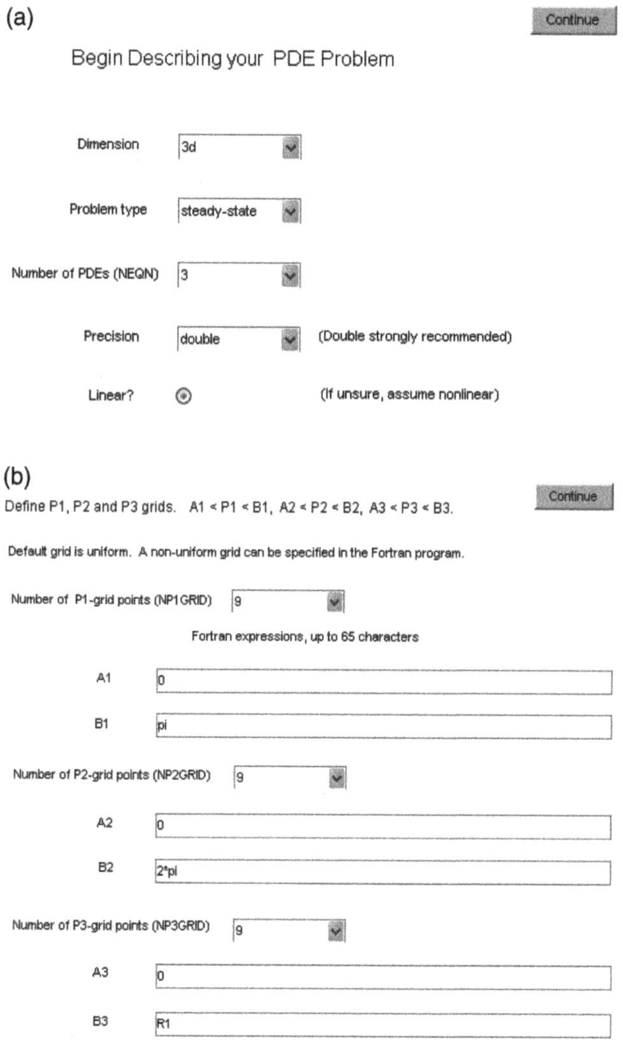

(a)

Continue

Begin Describing your PDE Problem

Dimension	3d
Problem type	steady-state
Number of PDEs (NEQN)	3
Precision	double (Double strongly recommended)
Linear?	⊙ (If unsure, assume nonlinear)

(b)

Define P1, P2 and P3 grids. A1 < P1 < B1, A2 < P2 < B2, A3 < P3 < B3.

Continue

Default grid is uniform. A non-uniform grid can be specified in the Fortran program.

Number of P1-grid points (NP1GRID) 9

Fortran expressions, up to 65 characters

A1 0

B1 pi

Number of P2-grid points (NP2GRID) 9

A2 0

B2 2*pi

Number of P3-grid points (NP3GRID) 9

A3 0

B3 R1

Figure I.6 (a and b) Two pages from GUI session.

Now enter FORTRAN expressions to define the PDE coefficients, which may be functions of

$$X,Y,U,Ux,Uy,V,Vx,Vy$$

They may also be functions of the initial triangle number KTRI and, in some cases, of the parameter T.

Recall that the PDEs have the form

$$d/dX*A1 + d/dY*B1 = F1$$
$$d/dX*A2 + d/dY*B2 = F2$$

```
  Do you want to write a FORTRAN block to define some parameters to be
  used in the definition of these coefficients?
  |---- Enter yes or no
->yes
  Remember to begin FORTRAN statements in column 7
  |-----7-----Input FORTRAN now (type blank line to terminate)-----------|
->      Rmu = 0.1
->      Rho = 22.0
->      P = -alpha*(Ux+Vy)
->      f1 = y
->      f2 = -x

  F1 =           (Press [RETURN] to default to 0)
  |----Enter constant or FORTRAN expression----------------------|
-> Rho*(U*Ux+V*Uy) - f1
  A1 =           (Press [RETURN] to default to 0)
  |----Enter constant or FORTRAN expression----------------------|
-> 2*Rmu*Ux - P
  B1 =           (Press [RETURN] to default to 0)
  |----Enter constant or FORTRAN expression----------------------|
-> Rmu*(Uy+Vx)
  F2 =           (Press [RETURN] to default to 0)
  |----Enter constant or FORTRAN expression----------------------|
-> Rho*(U*Vx+V*Vy) - f2
  A2 =           (Press [RETURN] to default to 0)
  |----Enter constant or FORTRAN expression----------------------|
-> Rmu*(Uy+Vx)
  B2 =           (Press [RETURN] to default to 0)
  |----Enter constant or FORTRAN expression----------------------|
-> 2*Rmu*Vy - P
```

Figure I.7 Portion of PDE2D interactive session.

offers fewer options than the interactive driver interface, but cannot handle completely general 2D regions (because it only supports the collocation methods), and has some other limitations, for example, it cannot handle more than eight simultaneous PDEs. The GUI basically makes relatively easy problems extremely easy and is very useful in the classroom, but users who need to solve more difficult problems should create their programs using the interactive driver. There is a video at www.pde2d.com that is a good tutorial on how to use the GUI and interactive driver to create PDE2D programs.

Both the GUI and interactive driver interfaces produce a Fortran90 program (*name*.f), which is compiled and linked to the PDE2D library, and run ("runpde2d *name*"). Figure I.8 shows part of the Fortran program generated by the interactive session of Figure I.7. Typically, users work through a GUI or interactive driver session once and then make minor corrections or changes to the model directly in this Fortran program. The questions seen in either session are repeated in the Fortran comments, which makes it easy to modify the program to change options you selected or input you provided during the GUI or interactive session and to change many other options that were selected for you by default (especially if you used the GUI); only if major changes to the model are required is it necessary to work through a new session. The fact that a Fortran program is generated gives PDE2D, despite the user-friendliness of the human interface, all the flexibility of Fortran: for example, you can write Fortran functions to define any PDE and boundary condition coefficients (see Example 3.4, for example) or write your own Fortran postprocessing code. However, users do *not* need to be too familiar with Fortran, usually all they need to know is how to form Fortran expressions, such as $(X + 3 * Y) * Ux$, which are similar to MATLAB expressions, and many of the same intrinsic functions (sqrt,abs,sin,...) are available. The program is generated automatically, and normally users only need to modify the Fortran expressions they provided during the GUI or interactive driver session and do not need to touch the more complicated portions of the program.

Here is a short summary of a few of the more important additional features of PDE2D. More detail on each of these is contained in the remaining chapters, and especially in Appendix B.

1) PDE2D not only produces its own graphical output in a PostScript file ("*name*.ps") but also will automatically produce a MATLAB program ("pde2d.m"), which can be run to produce certain MATLAB plots, and can be easily modified to generate any other graphics MATLAB can create or otherwise postprocess the PDE2D results.

2) For 2D problems, using the Galerkin option, a user-supplied initial triangulation can be refined adaptively or graded according to user-supplied specifications (see Figure I.1).

```
                          else
C#####################################################################
C      Now enter FORTRAN expressions to define the PDE coefficients, which  #
C      may be functions of                                                  #
C                                                                           #
C                          X,Y,U,Ux,Uy,V,Vx,Vy                              #
C                                                                           #
C      They may also be functions of the initial triangle number KTRI      #
C      and, in some cases, of the parameter T.                             #
C                                                                           #
C      Recall that the PDEs have the form                                   #
C                                                                           #
C                          d/dX*A1 + d/dY*B1 = F1                           #
C                          d/dX*A2 + d/dY*B2 = F2                           #
C                                                                           #
C#####################################################################
      Rmu = 0.1
      Rho = 22.0
      P = -alpha*(Ux+Vy)
      f1 = y
      f2 = -x
                  if (j8z.eq.0) then
      yd8z = 0.0
C                                                    F1 DEFINED
      if (i8z.eq.    1) yd8z =
     & Rho*(U*Ux+V*Uy) - f1
C                                                    A1 DEFINED
      if (i8z.eq.    2) yd8z =
     & 2*Rmu*Ux - P
C                                                    B1 DEFINED
      if (i8z.eq.    3) yd8z =
     & Rmu*(Uy+Vx)
C                                                    F2 DEFINED
      if (i8z.eq.    4) yd8z =
     & Rho*(U*Vx+V*Vy) - f2
C                                                    A2 DEFINED
      if (i8z.eq.    5) yd8z =
     & Rmu*(Uy+Vx)
C                                                    B2 DEFINED
      if (i8z.eq.    6) yd8z =
     & 2*Rmu*Vy - P
                  else
                  endif
                      endif
```

Figure I.8 Portion of Fortran program created by interactive session.

3) For 2D problems, using the Galerkin option, curved boundaries can be defined by parametric equations, or a cubic spline can be drawn through user-supplied boundary points.

4) The Galerkin finite element method options use up to 4th degree elements, thus up to $O(h^5)$ accuracy (h=element diameter). The collocation options use cubic elements and thus generally provide $O(h^4)$ accuracy.

5) Newton's method is used to solve the nonlinear algebraic equations, for nonlinear problems.
6) The shifted inverse power method is used to find a single eigenvalue, with associated eigenfunction, of an eigenvalue PDE. If a direct solver is used, the LU decomposition that is calculated the first iteration is used to make the other iterations run much faster. If all eigenvalues are desired (without eigenfunctions), a shifted QR iteration is used.
7) Adaptive time step control is available for time-dependent problems.
8) For 2D and 3D problems, there are several options to solve the large linear systems, including iterative methods, sparse direct solvers, and frontal methods. On multiprocessor systems with MPI message passing software, PDE2D also offers parallel linear system solver options (see Section I.4).
9) The GUI and interactive driver do a lot of error checking, and there are many runtime error checks also.
10) Accurate integrals of user-supplied functions of the solution and its derivatives can be calculated, in a 1D, 2D, or 3D region. Accurate boundary integrals can also be calculated over the boundary of the region.
11) The solution can be dumped and restarted exactly if the same finite element grid is used on restart or dumped on a regular grid and interpolated to restart using a different finite element grid. There is also a function that reads and interpolates the solution saved by another PDE2D program.

I.3 Accuracy

The collocation method's use of cubic elements, and the Galerkin method's use of up to 4th degree isoparametric elements, allows PDE2D to solve PDEs accurately even in curved regions. To document this, we solved:

$U_{xx} + U_{yy} = 2U - (16x + 36y + 26)e^{x+y}$ in the region of Figure I.9
with $U = 0$ on the outer (elliptical) boundary
and $U = (36 - 4x^2 - 9y^2)e^{x+y}$ on the rest of the boundary.

which has exact solution $U^* = (36 - 4x^2 - 9y^2)e^{x+y}$.

Table I.1 shows the L_1 relative error (the integral of $|U - U^*|$ divided by the integral of $|U^*|$) using the Galerkin method with fourth degree elements; the linear systems were solved by a sparse direct method, based on Harwell's MA37 code (Duff and Reid 1984). Notice that the relative error produced using 64 000 triangles, in only 45 seconds, is about $1.3 * 10^{-11}$. Of course, it will not be possible to achieve this much precision on more difficult problems.

When 64 000 *linear* (IDEG=1) elements were used, the relative error was $3.12 * 10^{-4}$, twenty million times larger than with 64 000 quartic (IDEG=4) elements. Of course there are fewer unknowns with linear elements, so a fairer comparison would be with the 4 000 quartic triangle test, which required the solution of a problem with exactly the same number of unknowns but produced

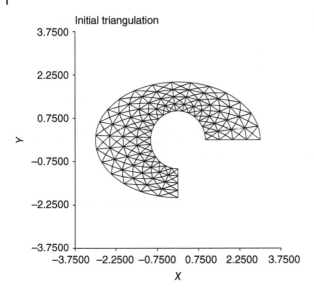

Figure I.9 Region used for accuracy tests.

Table I.1 Galerkin method (IDEG=4, ISOLVE=4).

Triangles	Unknowns	$\| U - U^* \|_1 / \| U^* \|_1$	CPU time (sec)	Memory (Mwords)
4 000	31 721	$8.07 * 10^{-9}$	0.8	2.8
16 000	127 441	$2.62 * 10^{-10}$	6.2	12.6
64 000	510 881	$1.31 * 10^{-11}$	45.7	57.0

Table I.2 Collocation method (ISOLVE=1).

Rectangles	Unknowns	$\| U - U^* \|_1 / \| U^* \|_1$	CPU time (s)	Memory (Mwords)
1 000	4 284	$1.06 * 10^{-6}$	1.4	0.8
4 000	16 564	$6.64 * 10^{-8}$	6.1	3.7
16 000	65 124	$4.06 * 10^{-9}$	28.5	18.3

an error 40 000 times smaller. When the region has a curved boundary, just using higher order elements does not improve on the asymptotic order ($O(h^2)$) of linear elements, unless isoparametric elements (see Section B.3) are used.

Table I.2 shows the relative errors when the collocation method was used. The PDE2D collocation method also produces high order ($O(h^4)$) accuracy in curved regions, because of the global transformation of coordinates used.

The CPU times given are for the entire problem, not just the linear system solution, on one processor of a UTEP Centos cluster.

I.4   Computer Time and Memory

When a linear steady-state PDE system is solved, PDE2D must solve one linear system, and when a nonlinear PDE system is solved, a linear system must be solved every Newton iteration. If a time-dependent problem is solved, a linear system must be solved every time step, and for eigenvalue problems, a linear system must be solved every shifted inverse power iteration. For 2D and 3D problems, the vast majority of the computer time and memory is used in solving these linear systems, so the PDE2D execution speed is very much dependent on the efficiency of the linear system solvers used.

For 2D problems, using the Galerkin method, there are five linear system solver options:

ISOLVE=1. A band solver, with a reverse Cuthill–McKee ordering of the unknowns, described in section 5.5 of Sewell (2015).

ISOLVE=2. An out-of-core band solver, called a frontal method, described in section 3.3 of Sewell (1985). This is essentially the same as the band solver, but it is slower and requires much less memory because the rows of the band matrix are written to disk after processing during the forward elimination and read back in when needed during the back substitution. Hence only a small portion of the band matrix is held in memory at any given time.

ISOLVE=3. A block diagonal preconditioned biconjugate gradient (Lanczos) iterative method (described in section 3.4 of Sewell (1985)). This is a generalization of the conjugate gradient method, designed to handle nonsymmetric linear systems. When multiple processors are available, this solver is "MPI enhanced," and the work is distributed over the available processors (see Section B.8).

ISOLVE=4. Harwell Library minimal degree sparse direct solver MA27 is used for symmetric problems (Duff and Reid 1983) and MA37 for nonsymmetric problems (Duff and Reid 1984). (A royalty was paid to Harwell Laboratory for use of these routines in PDE2D.)

ISOLVE=6. This is a parallel band solver, described in Section B.8, where the columns of the matrix are distributed over the available processors. This option is of course only available on multiple processor machines.

To test the relative efficiency of these solvers, we solved the 2D elasticity equations of Example 5.2 with 20 000 cubic (IDEG=3) triangles, and recorded the results in Table I.3. This system of PDEs is symmetric, and all of these solvers

Table I.3 2D Linear system solvers.

Unknowns = 180718, half bandwidth = 1621			
ISOLVE	NPES	CPU time (s) *min	Per-processor memory (Mwords)
1	1	342	294.8
2	1	411	3.2
3	1	256	29.6
3	2	154	29.6
3	4	*120	17.6
3	8	221	11.6
4	1	14	29.4
6	1	441	895.6
6	2	229	456.1
6	4	*136	246.4
6	8	152	119.8

take advantage of this to cut the time and memory. The tests were run on a UTEP Centos cluster of Intel XE processors, and execution times given are for the entire problem, not just the linear system solver.

As these results suggest, for 2D problems the sparse direct solver (ISOLVE=4) is nearly always faster than the other options, even than the parallel solvers (ISOLVE=3 and 6) when they use multiple processors (NPES = number of processing elements). The band solvers (ISOLVE=1,2,6) are somewhat more stable with respect to roundoff error than the sparse direct solver, however, so for ill-conditioned problems they may be needed.

For 3D problems (and 2D problems when the collocation method is used), PDE2D provides the following options to solve the linear systems $Ax = b$:

ISOLVE=1. A sparse direct solver, based on MA27, is used to solve the symmetric positive definite "normal" equations, $A^T A x = A^T b$ (See Section B.4.).
ISOLVE=2. A frontal method, which is just a band solver with out-of-core storage of the matrix. This option is slow but requires very little memory.
ISOLVE=3. A diagonal-preconditioned conjugate gradient iterative method (described in section 1.10 of Sewell (2014)) is used to solve the positive definite normal equations. This solver is also "MPI enhanced," as described in Section B.8.
ISOLVE=6. This is a parallel band solver, described in Section B.8, where the columns of the matrix are distributed over the available processors.

Table I.4 3D linear system solvers.

		Unknowns = 34560, half bandwidth = 4541	
ISOLVE	NPES	CPU time(s) *min	Per-processor memory (Mwords)
1	1	463	139.2
2	1	3337	18.8
3	1	1037	20.3
3	2	585	20.3
3	4	306	10.3
3	8	261	5.3
3	16	*198	2.8
6	1	1517	490.9
6	2	703	255.5
6	4	389	127.8
6	8	*213	64.0
6	16	333	32.1

Table I.4 shows the computer time and memory requirements when these four options were used to solve the linear system for the 3D elasticity Example 5.3, using a $12 \times 10 \times 12$ grid of tricubic Hermite elements (see Section B.4).

For 3D problems, the parallel options (ISOLVE=3 and 6) are often faster than the sparse direct solver (ISOLVE=1, the GUI default) and use less memory. Again, the band solvers (ISOLVE=2,6) are more stable numerically, so they may be needed if the problem is ill conditioned, which happens more frequently (see Example 6.4 for an example) for this sparse direct solver because it is applied to the normal equations rather than the original system $Ax = b$. (It runs much faster on the symmetric, positive definite, normal equations; see Section B.4.) As is normal for parallel numerical algorithms, as processors are added the times eventually quit improving and start to increase, though this happens later for larger problems.

For eigenvalue problems and time-dependent problems with NOUPDT= .TRUE., all the direct solvers save the LU decomposition calculated on the first shifted inverse power iteration or time step, to dramatically decrease the computer time on the second and subsequent steps. For example, when two iterations of the problem of Table I.4 were done, with NOUPDT=.TRUE., the first/second iterations took 463/7 seconds, respectively, with ISOLVE=1, 3337/33 seconds with ISOLVE=2, and 213/6 seconds with ISOLVE=6 on 8 processors. Thus, for such problems, the iterative solvers (ISOLVE=3) are not recommended.

Table I.5 All eigenvalues, 3D problem.

Unknowns= 12584	
NPES	Time (s)
1	53 574
2	31 847
4	22 786
8	13 544

PDE2D makes it easy for users to plug in their own linear system solvers and will even deliver the matrix already distributed over the processors, for a user-written MPI-based parallel linear system solver. This is a good way to generate a wide range of test problems for linear system solvers. The missing ISOLVE numbers (5 for the Galerkin solvers; 4,5 for the collocation solvers) are reserved for these user-supplied solvers. See Problem 7 of Chapter 5 for more on how to add your own solvers.

The most computationally expensive calculation done by PDE2D does not require solving a large linear system, however. It is the calculation of all the eigenvalues (ITYPE=4) of an eigenvalue problem, by applying the QR method to a generalized eigenvalue problem $A\mathbf{z} = \lambda B\mathbf{z}$. This requires a large amount of computer time, but it is parallelized on multiple processor systems, for 3D, 2D, and even 1D eigenvalue problems, as described in Section B.7. We tested this algorithm on the 3D eigenvalue equation of Problem 2a of Chapter 7 using a $11 \times 11 \times 13$ grid. The overall calculation scales fairly well up to 8 processors, as seen by the results in Table I.5.

Table I.6 shows the times, using the Galerkin method with 1000 cubic triangles, to find all eigenvalues of the symmetric 2D problem, $W_{xx} + W_{yy} + \lambda W = 0$ with $W = 0$ on the boundary of the square $0 \le x \le \pi, 0 \le y \le \pi$. The exact eigenvalues are $\lambda_{m,n} = m^2 + n^2$ for all positive integers m, n, and the first 150 eigenvalues are computed correctly to the nearest integer. When IDEG=3, PDE2D has to use the QR method to find the eigenvalues of a full matrix, but this is parallelized, and the calculation scales reasonably well up to four processors. When IDEG=−3, as discussed in Section B.7, the QR method is applied to a symmetric band matrix (with half bandwidth = 318), and there is a dramatic speed-up.

Finally, the Galerkin method with 1000 fourth degree elements was used to find all eigenvalues of the 1D problem $W_{xx} + \lambda W = 0$ with $W(0) = W(\pi) = 0$. The exact eigenvalues are $\lambda_n = n^2$ for all positive integers n, and the first 500 eigenvalues are computed correctly to the nearest integer. This is a symmetric problem, but when we set SYMM=.FALSE., PDE2D does not know it is symmetric and has to find all eigenvalues of a full matrix. Again, this is parallelized

Table I.6 All eigenvalues, 2D problem.

Unknowns= 4405		
NPES	IDEG	Time (s)
1	−3	67
1	3	2924
2	3	1519
4	3	873
8	3	795

Table I.7 All eigenvalues, 1D problem.

Unknowns= 3999		
NPES	SYMM	Time (s)
1	True	1.6
1	False	1243
2	False	739
4	False	499
8	False	433

and results are shown in Table I.7. When we reset SYMM to .TRUE., the QR method is applied to a symmetric band matrix, and there is an even more spectacular speed-up, because the half bandwidth of the matrix is now only 5.

I.5 Programming Hints

Here are some important programming hints for interactive driver users ("pde2d *name*"):

1) Double precision is strongly recommended, always.
2) If your problem is nonlinear, you will be asked:

```
If you don't want to read the FINE PRINT, enter 'yes' (strongly recommended).

++++++++++++++++ THE "FINE PRINT" (CAN USUALLY BE IGNORED) ++++++++++++++++
+ The partial derivatives of some of the PDE and boundary condition     +
+ coefficients are required by PDE2D.  These may be calculated          +
+ automatically using a finite difference approximation, or supplied    +
+ by the user.  Do you want them to be calculated automatically?        +
++++++++++++++++++++++++++++ END OF "FINE PRINT" ++++++++++++++++++++++++++
```

It is very unlikely that supplying these partial derivatives will be worth the extra effort, PDE2D can calculate them accurately for you, and in the worst case you will not get wrong answers, the convergence of the Newton iteration may just be slowed. So it is strongly recommended that you answer "yes" to this question, always. If you do decide to supply the partial derivatives manually, PDE2D will usually, but not always, be able to issue a warning if you make any mistakes.

3) When you are asked to supply parametric equations (as in (I.1)) to define the region, for a 2D or 3D collocation problem, if you want to supply your own equations, it is strongly recommended that you enter ITRANS=−3, not 3. It is very unlikely to be worth the extra effort to supply the partial derivatives yourself; PDE2D can calculate them accurately for you. Again, if you do decide to supply the partial derivatives manually (ITRANS=3), PDE2D will usually be able to issue a warning if you make any mistakes.

4) If you use the Galerkin method, for 1D or 2D problems, you will be asked if your problem is symmetric. It is always safe to say "no," and since symmetry will only cut the computer time and memory in half, it may or may not be worth the effort to decide if it is symmetric. One exception: If you are going to calculate all eigenvalues (ITYPE=4), then symmetry is much more important; see Section I.4. If you claim your problem is symmetric and it is not, PDE2D will usually, but not always, be able to issue a warning.

5) For time-dependent problems, if your problem is linear and you do not request an adaptive time step, you will be asked:

```
If you don't want to read the FINE PRINT, it is
safe (though possibly very inefficient) to enter 'no'.
```

```
++++++++++++++++ THE "FINE PRINT" (CAN USUALLY BE IGNORED) ++++++++++++++
+ If your time-dependent problem is linear with all PDE and boundary   +
+ condition coefficients independent of time except inhomogeneous      +
+ terms, then a large savings in execution time may be possible if     +
+ this is recognized (the LU decomposition computed on the first step  +
+ can be used on subsequent steps).  Is this the case for your         +
+ problem?  (Caution: if you answer 'yes' when you should not, you     +
+ will get incorrect results with no warning.)                         +
+++++++++++++++++++++++++++++++ END OF "FINE PRINT" ++++++++++++++++++++++++++++
```

As an example, if you are solving

$$c(x, y)U_t = \nabla \cdot [D(x, y)\nabla U] + f(x, y, t)$$

with boundary conditions

$$U = fb(x, y, t)$$

$$D(x, y)\frac{\partial U}{\partial n} = e(x, y)U + gb(x, y, t)$$

you can answer "yes," but if $c, D,$ or e depend on t, you cannot. Although it is always safe to answer "no," the savings in computer time may be huge if you answer "yes," and you are using a direct linear system solver (see Section I.4), so this question may be worth some consideration. Unfortunately, if you answer "yes" incorrectly, you will probably get wrong answers with no warning, so only answer "yes" if you are sure.

6) As mentioned earlier, if you need to make minor corrections or changes to your model, you do not need to go through another interactive driver or GUI session; you can make most changes directly to the Fortran program, even if you are not very familiar with Fortran. The questions you were asked in the GUI or interactive driver session are repeated in the comments; Figure I.8 illustrates how easy it is to change your PDE coefficients, for example. The comments can help you make many other changes without re-entering a new interactive session: for example, if you initially say you do not want to calculate any integrals of the solution but later change your mind, there are comments that make this easy.

   If you need to make *major* changes, for example, to increase the number of equations, or change from a steady-state model to a time-dependent model, or from a 2D problem to a 3D problem, you will need to start over with a new GUI or interactive driver session. However, notice that after an interactive session, there is a file "echo.out" in which everything you entered during the session is saved. If you entered some data during the first session that you don't want to have to re-enter – for example, the vertices of a complicated initial triangulation – you can rename "echo.out" to "pde2d.in" and remove everything but the data you don't want to re-enter. Then when prompted for this information, enter "#," and the interactive driver will read from "pde2d.in" until it encounters another "#" or an end of file.

7) For most of the programs in this book, it is not necessary to write any Fortran subprograms or blocks, only a few Fortran expressions defining the PDE coefficients and so forth. PDE2D users do have all the flexibility of Fortran at their disposal, however, and, occasionally, to take advantage of this flexibility, you may need to write a Fortran function or subroutine. For example, the user-written functions used in Problem 6 of Chapter 1 are repeated below for convenience:

```fortran
function D(x)
implicit double precision (a-h,o-z)
if (x < 0) then
   D = 10
else
   D = 1
endif
return
end
```

```
function True(x)
implicit double precision (a-h,o-z)
if (x < 0) then
   True = x**2/10.d0 - (5.9d0*x + 8)/21.d0
else if (x < 1) then
   True = x**2 - (59*x + 8)/21.d0
else
   True = x**2 - (17*x + 50)/21.d0
endif
return
end
```

Here are a few comments on these functions to provide tips for users less familiar with Fortran:

a) Note the use of "implicit double precision (a-h,o-z)" in both functions. Assuming you request double precision, which you always should, this ensures that function names and argument types agree with those in the PDE2D calling program (at least if the argument names match), which uses the convention that names starting with letters I,J,K,L,M,N are typed integer, all others are typed double precision. It is essential that argument types match those in the calling program. If you request single precision, you could use "implicit real (a-h,o-z)," but this is unnecessary since that is the Fortran default.

b) Fortran is *not* case sensitive, "true" is the same as "TRUE" or "True."

c) Notice that x^2 is written as x**2 in Fortran.

d) For full precision, it is safest to write all double precision constants using the "d0" suffix, though this is usually unnecessary. Be especially careful with integer division: (1/3)=0, while (1.d0/3.d0) and (1/3.d0) = 0.333333333333333d0.

e) Fixed format Fortran is used, so statements must be in columns 7–72 except for statement numbers, which should be in columns 1–5.

1

The Damped Spring and Pendulum Problems

1.1 Derivation of the Damped Spring and Pendulum Equations

In this chapter we present some simple ordinary differential equation problems to give students a chance to become familiar with PDE2D before proceeding to more difficult problems in later chapters.

Like many second-order differential equations, the equations used to model the damped spring and pendulum are derived using Newton's second law: Mass times acceleration equals the force acting on the mass.

Suppose a weight of mass m hangs from the ceiling on a spring. We will let $y(t)$ be the height of this weight, with $y = 0$ taken as its height when stationary. Then we will consider three forces acting on this mass: The force of the spring itself will be approximately proportional to the displacement from equilibrium and in the opposite direction, $-ky$; a force of friction (perhaps due to the surrounding air or liquid, or the spring itself) approximately proportional to the velocity of the mass and in the opposite direction, $-by'$; and an additional external force, $f(t)$, which may be caused by some outside agent such as a magnetic force. Thus, according to Newton's second law

$$my'' = -ky - by' + f(t) \tag{1.1}$$

Why didn't we include the force of gravity on this mass? Well, actually we did, by taking $y = 0$ to be the height when there is no force acting on the mass *other than gravity*, see Problem 1.

Since this is a second-order equation, we will need two initial conditions, we need to specify the initial position $y(0)$ and the initial velocity $y'(0)$.

The kinetic energy associated with the spring mass is $\frac{1}{2}m(y')^2$. Its potential energy is $\frac{1}{2}ky^2$, because that is how much energy is required to move it a distance y from equilibrium, against an opposing force increasing linearly from 0 to ky, thus against an average opposing force of $\frac{1}{2}ky$. So the total (kinetic plus

Solving Partial Differential Equation Applications with PDE2D, First Edition. Granville Sewell.
© 2018 John Wiley & Sons, Inc. Published 2018 by John Wiley & Sons, Inc.

potential) energy of the mass is $E(t) = \frac{1}{2}m(y')^2 + \frac{1}{2}ky^2$. Then if there is no external force adding energy to the system, $E'(t) = my'y'' + kyy' = y'(my'' + ky) = -b(y')^2$, so the total energy is constant if there is no friction and decreases if there is (until a steady state has been reached, $y' = 0$).

Now let's consider a mass m suspended by a rigid wire of length L fastened at the origin. Let $(x(t), y(t))$ be the position of the mass, and we assume that the only forces acting on this mass are the force of gravity, $(0, -mg)$, and the wire itself, which exerts a force of magnitude equal to the tension in the wire, $\lambda(t)$, in the direction of the unit vector toward the origin, $(-x(t)/L, -y(t)/L)$. Newton's second law gives us the pendulum equations:

$$mx'' = -\lambda x/L$$
$$my'' = -\lambda y/L - mg$$

If we break the two second-order equations into first-order equations by defining $u \equiv x', v \equiv y'$ and add the constraint that $x^2 + y^2 = L^2$, we get five equations for the five unknown functions x, y, u, v, λ:

$$x' = u$$
$$y' = v$$
$$mu' = -\lambda \frac{x}{L}$$
$$mv' = -\lambda \frac{y}{L} - mg$$
$$0 = x^2 + y^2 - L^2 \tag{1.2}$$

The total energy of the pendulum is the kinetic energy plus the potential energy:

$$E(t) = \frac{1}{2}m[(x')^2 + (y')^2] + mgy \tag{1.3}$$

This total energy should remain constant, since

$$E'(t) = mx'x'' + my'y'' + mgy' = x'\left(-\lambda\frac{x}{L}\right) + y'\left(-\lambda\frac{y}{L} - mg\right) + mgy'$$
$$= -\frac{\lambda}{L}(xx' + yy') = -\frac{\lambda}{2L}(x^2 + y^2)' = 0$$

Problem (1.2) is a differential-algebraic system because the last equation has no derivatives. In fact it is given as an example of an "index 3" system in the documentation for IMSL Library differential-algebraic system solver DAESL (IMSL, Inc. 2010). The documentation says it cannot be solved by DAESL until the last equation has been differentiated twice by hand. So this simple-looking system is not quite as easy to solve as it appears.

1.2   Damped Spring and Pendulum Examples

Example 1.1 (Damped spring) We first solve the damped spring equation (1.1) using PDE2D, with $m = 1, b = 2, k = 101, f(t) = 0$, and initial conditions $y(0) = 0, y'(0) = 10$. This is a "0D" (no space variables), time-dependent problem, but PDE2D only handles first derivatives in time, so it must be written as a system of two first-order ordinary differential equations:

$$y' = v$$
$$mv' = -ky - bv$$

A plot of $y(t)$ is shown in Figure 1.1. Notice the solution dies out, due to the frictional force, as it oscillates. The exact solution is $y(t) = e^{-t}\sin(10t)$.

Example 1.2 (Pendulum) We next solve the pendulum equations (1.2), with $m = 4, L = 1, g = 32$, and initial conditions $x(0) = L, y(0) = 0, u(0) = 0, v(0) = 0, \lambda(0) = 0$, which means we start with the pendulum horizontal, 90° to the right of its low point. This "index 3" differential-algebraic system is indeed difficult to solve for most finite difference methods: When we asked PDE2D to choose a time step adaptively (ADAPT=.TRUE.), it gave a warning every step that it had taken the minimum step size and still could not satisfy the default error tolerance, and when we requested a very small constant time step, using a second-order Adam's implicit method (CRANKN=.TRUE.), there

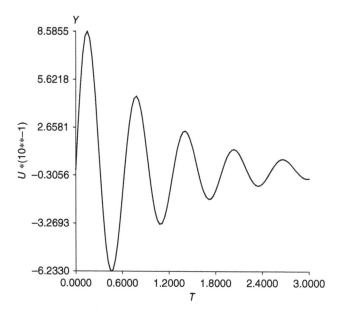

Figure 1.1 Damped spring oscillation and decay, Example 1.1.

were huge oscillations. But with ADAPT=.FALSE.,CRANKN=.FALSE., and NSTEPS=1000, which means 1000 steps are taken with a simple backward Euler method, we got the reasonable results shown in Figure 1.2. The backward Euler method is actually ideally suited for differential-algebraic systems (except of course that it is only $O(dt)$ accurate!) and should always be selected when solving mixed time-dependent and steady-state PDEs with PDE2D.[1]

Figure 1.2a shows a pendulum whose oscillations are dying out, but there was no damping included in model (1.2); the total energy in this model should be constant. The damping is due to the low-order backward Euler method's well-known tendency to dissipate energy, and rerunning with NSTEPS=100 000 produced much less damping, as seen in Figure 1.2b.

1.3 Problems

1 (Effect of gravity on spring) Suppose we add gravity as an external force to Eq. (1.1), that is, suppose $my'' = -ky - by' - mg + f(t)$. That will displace the mass downward at rest a distance mg/k (force divided by the spring constant.) Now define $z \equiv y + mg/k$, so the new rest position $y = -mg/k$ corresponds to $z = 0$. Show that Eq. (1.1) becomes $mz'' = -kz - bz' + f(t)$, so this equation really does take gravity into account, as long as $z = 0$ ($y = 0$ in the original equation) means the rest height with gravity on.

2 (Resonance in spring) Solve Eq. (1.1) with $m = 1, b = 0, k = 16, f(t) = sin(\omega t), y(0) = 0, y'(0) = 0$. Solve first with $\omega = 3$, then with $\omega = 4$, and plot the solution $y(t)$ as a function of time (see Figure 1.3). Use the PDE2D GUI ("pde2d_gui name") to create your program and "runpde2d name" to run the program.

   Equation (1.1), with external force $f(t) = sin(\omega t)$, can be solved analytically. If $b^2 < 4mk$, the general solution is

$$y(t) = C_1 e^{-\alpha t} sin(\beta t) + C_2 e^{-\alpha t} cos(\beta t) + \frac{sin(\omega t - \phi)}{\sqrt{(k - m\omega^2)^2 + (b\omega)^2}}$$

where $\alpha = b/(2m), \beta = \sqrt{4mk - b^2}/(2m), \phi = tan^{-1}[b\omega/(k - m\omega^2)]$. From this we see that if the frictional coefficient b is small, and ω is close to the resonant frequency $\sqrt{k/m}$ (= 4 in this problem), the solution will have an oscillating term of frequency ω, with a large amplitude. If, as in your problem, $b = 0$ and $\omega = \sqrt{k/m}$, the denominator in the analytical solution given above is zero, so it is no longer a valid solution. What is the general solution then?

---

1 For a system of equations $c_i u_i' = f_i(t, u_1, ..., u_N)$, backward Euler is $c_i[u_i^{n+1} - u_i^n]/dt = f_i(t^{n+1}, u_1^{n+1}, ..., u_N^{n+1})$. If the i-th equation is algebraic, $c_i = 0$, and this becomes simply $0 = f_i(t^{n+1}, u_1^{n+1}, ..., u_N^{n+1})$, which means the algebraic equation is enforced exactly every step.

(a)

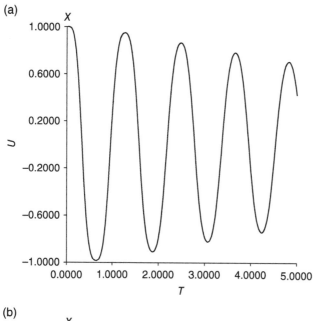

(b)

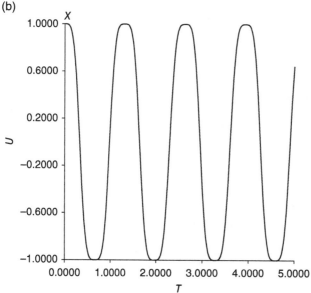

Figure 1.2 Plot of $X(t)$, pendulum Example 1.2. (a) NSTEPS=1000 and (b) NSTEPS=100 000.

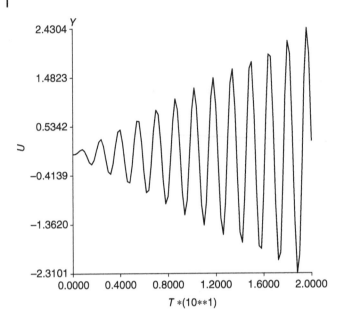

Figure 1.3 Resonance when $\omega = 4$, Problem 2.

3 (Looping pendulum) Use the GUI to recreate the pendulum Example 1.2. Do not request an adaptive time step. The GUI will set CRANKN=. TRUE. by default; you need to change this to .FALSE. with an editor (if you use the interactive driver, it will give you a choice). The GUI will set up time plots for all five variables; replace (using one of the elements of UPRINT(*) in PMOD8Z) one of these with a plot of the total energy (1.3), or else output the "integral" (for 0D problems, integral = value) of the total energy. You will see that the energy continually decreases, but with a large enough value for NSTEPS, you can slow this artificial damping.

Now instead of just dropping the pendulum from its horizontal position, give it a shove downward, that is, reset $v(0) = y'(0) = -7.5$. The pendulum will now go *past* the horizontal position on the other side and start upward, but as seen in Figure 1.4, it will not have enough energy to reach the top. (The tension can be negative because the mass is attached to a wire, not a string!) Rerun with a little stronger downward shove, and the pendulum will start looping, that is, $X(t)$ will keep increasing past zero, instead of turning back. Using the total energy formula (1.3), you can calculate exactly how big $v(0)$ needs to be for the pendulum to have enough energy to reach the top, assuming no artificial energy dissipation.

Increase $|v(0)|$ even further, until the tension remains positive even when $y > 0$ due to the centrifugal force.

(a)

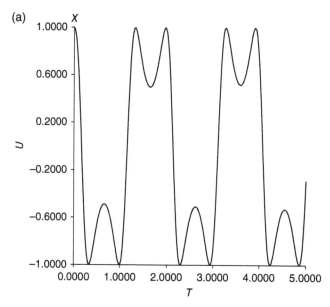

(b)

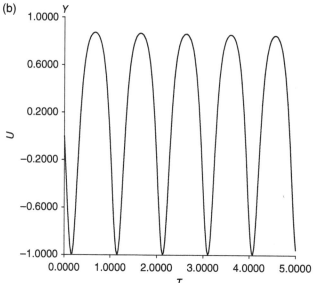

Figure 1.4 Pendulum Problem 3, $V(0) = -7.5$. (a) $X(t)$ and (b) $Y(t)$.

4 (Complex differential equation) Solve the differential equation $\frac{du}{dt} = u^{11}$, $u(0) = 1 + i$. Because of the complex initial condition, the solution will be complex, and so this must be broken into two real differential equations by defining $u = UR + i * UI$, $\frac{dUR}{dt} + i\frac{dUI}{dt} = (UR + i * UI)^{11}$, $UR(0) = 1$, $UI(0) = 1$. You could multiply out $(UR + i * UI)^{11}$ by hand and separate into real and imaginary parts, but it is much easier to let Fortran intrinsic functions do the work for you. Write the two equations as

$$\frac{dUR}{dt} = DREAL(DCMPLX(UR, UI) ** 11)$$

$$\frac{dUI}{dt} = DIMAG(DCMPLX(UR, UI) ** 11)$$

Compare your PDE2D solution with the exact solution, which is [2]

$$u = Re^{i\theta}$$
$$UR = R\,cos(\theta)$$
$$UI = R\,sin(\theta)$$
$$\text{where} \quad R = \sqrt{2}/[1 + (320t)^2]^{\frac{1}{20}}$$
$$\theta = \frac{\pi}{4} + \frac{1}{10} tan^{-1}(320t)$$

You may output the "integral" (=value) of, say, $|UR - R\,cos(\theta)| + |UI - R\,sin(\theta)|$ to measure the error.

This technique for solving complex PDEs has been used to solve some difficult complex applications with PDE2D (Alidoust, Sewell, and Linder 2012; Alidoust and Linder 2013). See also Example 7.2.

5 (Finding eigenvalues) In this section we have solved some ordinary differential equations systems, which are considered by PDE2D to be 0D time-dependent problems. PDE2D also solves 0D steady-state problems, that is, linear or nonlinear algebraic systems, and 0D eigenvalue problems, that is, matrix or generalized matrix eigenvalue problems. Use the GUI to find the eigenvalue closest to $p = 5$ and the associated eigenvector of the algebraic eigenvalue problem:

$$-5 * X1 + 7 * X2 + 3 * X3 + 4 * X4 - 8 * X5 = \lambda * X1$$
$$5 * X1 + 8 * X2 + 3 * X3 + 6 * X4 + 8 * X5 = \lambda * X2$$
$$3 * X1 - 7 * X2 + 9 * X3 - 4 * X4 + 5 * X5 = \lambda * X3$$
$$-3 * X1 + 4 * X3 + 5 * X4 + 3 * X5 = \lambda * X4$$
$$7 * X1 + 4 * X2 + 5 * X3 + 9 * X4 + 5 * X5 = \lambda * X5$$

---

2 Using separation of variables and the initial condition, it is easy to find that $u^{10} = 32(-320t + i)$ $/[1 + (320t)^2]$. But since there are 10 *different* 10th roots of the complex number on the right-hand side, you have to be careful to get the one that really satisfies the initial condition!

The inverse power method will be used to find the eigenvalue closest to p (called EV0R in the PDE2D program), and the associated eigenvector can be output from POSTPR. Then reset ITYPE to 4 in the Fortran program, and PDE2D will use the shifted QR method to find all five eigenvalues without eigenvectors. (Answer: $13.1406621 \pm 4.9368807\,i, -4.5805667 \pm 6.9420509\,i, 4.8798093$)

6 (Boundary value problem) Consider the boundary value problem:

$$(D(x)W_x)_x = 2 + 2\delta(x - 1)$$
$$W(-1) = 0$$
$$W(2) = 0$$

where $D(x)$ is discontinuous: $D(x) = 10$ for $x < 0$ and $D(x) = 1$ for $x \geq 0$, and δ is the Dirac delta function. Although this is an ordinary differential equation, it is considered by PDE2D to be a "1D" steady-state problem. Because of the discontinuous $D(x)$ and the Dirac delta, the Galerkin method must be used, so you need to create your program using the interactive driver ("pde2d *name*"). You can write Fortran functions D(x) and True(x) at the end of your PDE2D program that may look like this:

```
function D(x)
implicit double precision (a-h,o-z)
if (x < 0) then
   D = 10
else
   D = 1
endif
return
end

function True(x)
implicit double precision (a-h,o-z)
if (x < 0) then
   True = x**2/10.d0 - (5.9d0*x + 8)/21.d0
else if (x < 1) then
   True = x**2 - (59*x + 8)/21.d0
else
   True = x**2 - (17*x + 50)/21.d0
endif
return
end
```

Function "True" returns the known exact solution of this problem. You should request that the integral of "abs(W-true(x))" be calculated to

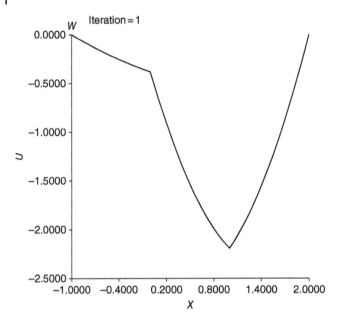

Figure 1.5 Boundary value Problem 6 solution.

measure the error. You will find that if you use second-degree elements or higher (IDEG=2,3, or 4), the error is down to roundoff error (since the solution is a piecewise quadratic polynomial) even if you only have three elements (NXGRID=4), provided you put a gridpoint at $x = 0$, where $D(x)$ and W_x are discontinuous, and another at $x = 1$, where the Dirac delta function is infinite, making W_x discontinuous there also (Figure 1.5). In Section I.5 some hints for programmers less familiar with Fortran are provided, which reference the above functions.

As an alternative to adding Fortran functions at the end of your program, you can define variables "D" and "True" by adding the above IF blocks in-line, when prompted by the interactive driver or later with an editor.

2

Beam and Plate Bending

2.1 Derivation of Beam Bending Equation

In this chapter we consider a rigid (1D) beam and a rigid (2D) plate, with external vertical forces.

To derive the beam bending equation, suppose $u(x)$ is the height of the beam with $u(0) = g_0, u'(0) = h_0, u(L) = g_1, u'(L) = h_1$, which minimizes the energy[1]

$$E(u) \equiv \int_0^L \frac{1}{2}D[u'']^2 - uq \, dx,$$

where $D(x)$ is the bending stiffness and $q(x)$ is an external vertical force (per unit length). Then if $e(x)$ is any smooth function with $e(0) = e'(0) = e(L) = e'(L) = 0$ on the boundary, $E(u + \alpha e) \geq E(u)$ for any α, thus

$$f(\alpha) \equiv E(u + \alpha e) = \int_0^L \frac{1}{2}D[u'' + \alpha e'']^2 - [u + \alpha e]q \, dx$$

should have a minimum at $\alpha = 0$ and so $\frac{df}{d\alpha}(0)$ should be zero. But

$$\frac{df}{d\alpha}(\alpha) = \int_0^L D[u'' + \alpha e'']e'' - eq \, dx$$

so

$$\frac{df}{d\alpha}(0) = \int_0^L Du''e'' - eq \, dx = \int_0^L [(Du''e')' - [(Du'')'e]' + (Du'')''e - eq] \, dx$$

$$= \int_0^L [(Du'')''e - eq] \, dx + D(L)u''(L)e'(L) - D(0)u''(0)e'(0)$$

$$- (D(L)u''(L))'e(L) + (D(0)u''(0))'e(0)$$

---

1 Although a rigorous derivation of this formula for the energy of a beam is beyond the scope of this book, it may be made more intuitive by noting that the potential energy of the beam should decrease by $uq \, dx$ if a segment dx of the beam is displaced a distance u in the direction of the force $q \, dx$, and it also should decrease when $(u'')^2$ decreases, since the second derivative measures curvature and a rigid beam resists being curved.

Solving Partial Differential Equation Applications with PDE2D, First Edition. Granville Sewell.
© 2018 John Wiley & Sons, Inc. Published 2018 by John Wiley & Sons, Inc.

Here we have essentially done two integrations by parts. Then, since $e(0) = e'(0) = e(L) = e'(L) = 0$,

$$\frac{df}{d\alpha}(0) = \int_0^L e[(Du'')'' - q] \, dx = 0$$

Since e is an arbitrary smooth function in the interior of $(0, L)$, we conclude that the factor multiplying e must be zero everywhere, which gives the fourth-order beam equation:

$$(D(x)u'')'' = q(x) \tag{2.1}$$

Note that the boundary conditions $u'' = 0, (Du'')' = 0$ at either end will also make the boundary terms disappear, so these are also possible boundary conditions.

The boundary conditions $u = g, u' = h$ are called "clamped" boundary conditions because they model a beam whose height is fixed and whose slope is "clamped" at an endpoint. "Simply supported" boundary conditions, $u = g, u'' = 0$, model a beam whose height is fixed at an endpoint, but whose slope is not clamped. If neither the height nor the slope are fixed, the boundary conditions are "free and unsupported": $u'' = 0, (Du'')' = 0$.

2.2 Derivation of Plate Bending Equation

The derivation of the plate equation is very similar. Now, suppose $u(x, y)$ is the height of the plate with $u = g(x, y), \frac{\partial u}{\partial n} = h(x, y)$ on the boundary $\partial\Omega$ of Ω that minimizes the energy:

$$E(u) \equiv \iint_\Omega \frac{1}{2} D[\nabla^2 u]^2 - uq \, dA,$$

where $D(x, y)$ is the bending stiffness and $q(x, y)$ is an external vertical force (per unit area). Then if $e(x, y)$ is any smooth function with $e = 0, \frac{\partial e}{\partial n} = 0$ on the boundary, $E(u + \alpha e) \geq E(u)$ for any α, thus

$$f(\alpha) \equiv E(u + \alpha e) = \iint_\Omega \frac{1}{2} D[\nabla^2 u + \alpha \nabla^2 e]^2 - [u + \alpha e]q \, dA$$

should have a minimum at $\alpha = 0$ and so $\frac{df}{d\alpha}(0)$ should be zero. But

$$\frac{df}{d\alpha}(\alpha) = \iint_\Omega D[\nabla^2 u + \alpha \nabla^2 e]\nabla^2 e - eq \, dA$$

so

$$\frac{df}{d\alpha}(0) = \iint_\Omega D\nabla^2 u \nabla^2 e - eq \; dA$$

$$= \iint_\Omega \nabla \cdot (D\nabla^2 u \nabla e) - \nabla(D\nabla^2 u) \cdot \nabla e - eq \; dA$$

$$= \int_{\partial\Omega} D\nabla^2 u \frac{\partial e}{\partial n} ds + \iint_\Omega -\nabla \cdot (e\nabla(D\nabla^2 u)) + e\nabla^2(D\nabla^2 u) - eq \; dA$$

$$\int_{\partial\Omega} \left(D\nabla^2 u \frac{\partial e}{\partial n} - e\frac{\partial(D\nabla^2 u)}{\partial n} \right) ds + \iint_\Omega e\nabla^2(D\nabla^2 u) - eq \; dA$$

Here we have done two (multivariate) integrations by parts (formula (A.3)). Then, since $e = 0, \frac{\partial e}{\partial n} = 0$ on $\partial\Omega$:

$$\frac{df}{d\alpha}(0) = \iint_\Omega e(\nabla^2(D\nabla^2 u) - q) \; dA = 0$$

Since e is an arbitrary smooth function in the interior of Ω, we conclude that the factor multiplying e must be zero everywhere, which gives the fourth-order plate bending equation:

$$\nabla^2(D(x, y)\nabla^2 u) = q(x, y) \tag{2.2}$$

Note that the free and unsupported boundary conditions $\nabla^2 u = 0, \frac{\partial(D\nabla^2 u)}{\partial n} = 0$ will also make the boundary integral terms disappear, so these are also possible boundary conditions, in addition to the clamped boundary conditions $u = g, \frac{\partial u}{\partial n} = h$ and the simply supported conditions $u = g, \nabla^2 u = 0$.

2.3 Beam and Plate Examples

Example 2.1 (Beam bending) We solve the beam bending problem (2.1) with $D(x) = 10, q(x) = -1, L = 2.5$, so there is a constant downward force on the beam, namely, the weight of the beam itself. We assume the beam is clamped at the left end, $u(0) = 0, u'(0) = 0$, and free and unsupported at the right end, $u''(L) = 0, u'''(L) = 0$.

Equation (2.1) is fourth order, so it must be written as two second-order equations for PDE2D:

$$\frac{d^2 u}{dx^2} = \frac{M}{D(x)}, \qquad u(0) = 0, u'(0) = 0$$

$$\frac{d^2 M}{dx^2} = q(x), \qquad M(L) = 0, M'(L) = 0$$

where $M(x)$ is the beam moment.

This problem has an analytical solution, $u_{true}(x) = -0.1(x^4/24 - Lx^3/6 + L^2x^2/4)$, and with $NXGRID = 10$ gridpoints, using the collocation method, PDE2D calculates the L_1 norm of the error, that is, the integral of $|u - u_{true}|$, as about 10^{-6}. When the Galerkin method is used, with only $NXGRID = 3$ fourth-degree elements, the integral is close to roundoff error, 10^{-11}, because the exact solution is a fourth-degree polynomial.

Example 2.2 (Plate with point loading) Next we solved the plate bending problem (2.2) in the unit square, with $D(x,y) = 1$ and $q(x,y) = -\delta(x-0.5, y-0.5)$, where δ is the Dirac delta function, so there is a load only at the midpoint of the square plate. We have to use the Galerkin method because of the point loading. Again, we have to reduce this fourth-order equation to a system of two second-order equations:

$$\nabla^2 u = M$$
$$\nabla^2 M = -\delta(x - 0.5, y - 0.5)$$

We assume clamped boundary conditions ($u = \frac{\partial u}{\partial n} = 0$) at $x = 0$ and $y = 0$ and simply supported boundary conditions ($u = M = 0$) at $x = 1$ and $y = 1$.

Using a property of the delta function, and two integrations by parts (formula (A.3)), and the fact that the boundary integrals disappear because of the boundary conditions, we see that

$$u(0.5, 0.5) = \iint_\Omega u\, \delta(x - 0.5, y - 0.5)\, dA = -\iint_\Omega u \nabla^2 M\, dA =$$
$$-\int_{\partial\Omega} u\frac{\partial M}{\partial n}\, ds + \int_{\partial\Omega} M\frac{\partial u}{\partial n}\, ds - \iint_\Omega M\nabla^2 u\, dA = \iint_\Omega -M^2\, dA$$

Thus to check the PDE2D solution, we compare the value of the solution at the midpoint with the PDE2D-calculated integral of $-M^2$, and the agreement is very good (-0.007426). A surface plot of the plate is shown in Figure 2.1. (Note: The minimum value as shown in this plot does not occur exactly at the midpoint.)

2.4 Problems

1 (Beam problem, collocation method) Solve the beam problem (2.1) with $D(x) = 1 + (x-1)^2$, $q(x) = -e^{-(x-1)^2}$, $L = 2$, and clamped boundary conditions, $U = 0, U' = 0$ at both ends. Note that parameters and variables must begin with a letter in the range $A - H$ or $O - Z$; otherwise, they will be integers, so, for example, do not use M for the moment, use, say, RM. Output the integral of $U(x)$ over the beam, and plot the beam height (see Figure 2.2).

Figure 2.1 Example 2.2 solution.

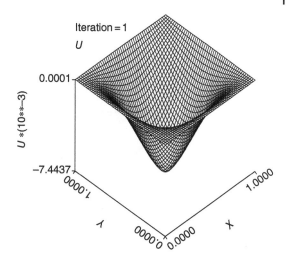

Figure 2.2 Solution for beam Problems 1 and 2.

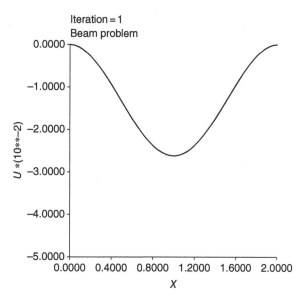

Use the PDE2D GUI ("pde2d_gui *name*"); that means the collocation method will be used.

2 (Beam problem, Galerkin method) Repeat Problem 1 using the Galerkin method. For this you must use the interactive driver ("pde2d *name*"). Notice that the format for the PDEs and (especially) the boundary conditions are very different than for the collocation method, so pay close attention to the documentation, especially to the hint on how to handle "mixed" (e.g. one

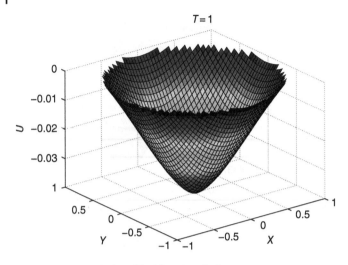

Figure 2.3 MATLAB plot of Problem 3a solution.

fixed, one free) boundary conditions, which must be treated as free conditions. If $A1 = U_x, A2 = RM_x$, you can set $GB1 = 0, GB2 = \text{zero}(U)$, which means $U_x N_x = 0, RM_x N_x = \beta U$, where β is a large number. (Setting $GB2 = \text{zero}(U)$ is exactly the same as $GB2 = \beta U$, except that PDE2D chooses an appropriate large β for you.) Again, output the integral of $U(x)$ to confirm you are getting the same solution as using the collocation method.

3 (Round plate with point loading, Galerkin method)
a) Solve the plate bending problem (2.2) using the PDE2D Galerkin method, with $D(x, y) = 1, q(x, y) = -\delta(x, y)$, in the unit disk; thus, there is a point load at the center of the disk. Use simply supported boundary conditions, $u = M = 0$, where $M \equiv \nabla^2 u$. Use initial triangulation generation option INTRI=2, with X=P*cos(Q), Y=P*sin(Q). Then IPARC(1) = 0 since P=0 is just a point, and IQARC(1)=IQARC(2)=1000, since the two edges Q=0 and Q= 2π coincide. Make a surface or contour plot of u (see Figure 2.3).

   As a check on the solution, let PDE2D calculate the boundary integral of $\frac{\partial M}{\partial n}$. Integrate both sides of (2.2), and use the divergence theorem to evaluate the integral of $\nabla^2 M$ (formula (A.1b)), and you will see what this integral should equal.
b) Find the analytical solution to Problem 3a as follows. Clearly u and M will be functions of $r = \sqrt{x^2 + y^2}$ only, so we can write the Laplacians in (2.2) in polar coordinate form as (cf. Problem 4d of Chapter 5):

$$\frac{1}{r}(ru_r)_r = M$$

$$\frac{1}{r}(rM_r)_r = -\delta(x,y)$$

Since $\delta(x,y) = 0$ for $0 < r \le 1, (rM_r)_r = 0$, so $M = C\,ln(r) + D$. The boundary condition $M(1) = 0$ implies $D = 0$, so $M = C\,ln(r)$. To find C, use

$$-1 = \iint_R -\delta(x,y)\,dx\,dy = \iint_R \nabla^2 M\,dx\,dy =$$

$$\int_{\partial R} \frac{\partial M}{\partial n}\,ds = 2\pi r\,M_r = 2\pi\,C.$$

where R is the disk $x^2 + y^2 \le r^2$, for arbitrary $0 < r \le 1$, and we have used the fact that $\frac{\partial M}{\partial n} = M_r$ is constant on the boundary ∂R and used the divergence theorem (Appendix A, formula (A.1b)) to replace the area integral with a boundary integral.

Now continue with $(ru_r)_r = rM$, integrate this twice and find the two constants from the boundary conditions $u(1) = 0$ and $u(0) < \infty$. (Hint: $\int r\,ln(r)\,dr = \frac{r^2 ln(r)}{2} - \frac{r^2}{4}$.) Edit your program from Problem 3a to calculate the integral of the absolute value of the error (don't forget to set NINT=1), and rerun.

4 (Round plate with point loading, collocation method) The collocation method does not allow point loading, but you can approximate $\delta(x,y)$ by the bivariate normal function $e^{-(x^2+y^2)/(2\sigma^2)}/(2\pi\sigma^2)$, when the standard deviation σ is small, because then this function has a very large, sharp peak near the origin, and an integral of one. Resolve Problem 3a using the collocation method, with $\delta(x,y)$ replaced by this approximation, and use ITRANS=1, that is, X=p1*cos(p2), Y=p1*sin(p2). At p1=0 there is no boundary condition (enter NONE), and there are periodic boundary conditions at p2= 0, 2π. Now it is essential to use a nonuniform grid, with several gridlines between $p1 = 0$ and, say, $p1 = 5\sigma$.

Again compute the boundary integral of $\frac{\partial M}{\partial n}$ and the integral of the absolute value of the error, using your exact solution from Problem 3b.

3

Diffusion and Heat Conduction

3.1 Derivation of Diffusion Equation

The diffusion equation is one of the most important PDE applications, so let's see how it is derived. We let $C(x, y, z, t)$ be the density (mass per unit volume) of a diffusing substance X, and let S be any small subregion of the region where diffusion is occurring. Then the total mass of X within the subregion is $\iiint_S C\, dV$, and

$$\frac{d}{dt} \iiint_S C\, dV = \iint_{\partial S} \mathbf{J} \cdot (-\mathbf{n}) \quad dA + \iiint_S q\, dV$$

This equation says that the rate of change of the total mass of X in S is equal to the net rate at which X is entering from outside plus the net rate at which X is being created internally due to sources and sinks. The flux vector $\mathbf{J}$ represents the net flow of X, in mass per unit area per unit time due to diffusion or convection and can be thought of as the density C times the average velocity of the particles. Thus the dot product of the flux vector with the unit inward normal vector $(-\mathbf{n})$ is the density times the component of the average velocity inward through the boundary, so the above boundary integral gives the net mass per unit time entering the subregion S. $q(x, y, z, t)$ is the net rate, in mass per unit volume per unit time, which X is being created within S, so the second integral on the right gives the total mass per unit time created (or destroyed) inside S.

Now using the divergence theorem (formula (A.1)), we see that

$$\iiint_S [C_t + \nabla \cdot \mathbf{J} - q]\, dV = 0$$

Since S is an arbitrary, small subregion, this means that the integrand must be zero everywhere.

Now Fick's law says that the flux due to (isotropic) diffusion is in the direction of most rapid decrease in density $(-\nabla C)$, and its magnitude is proportional to the rate of decrease in that direction, $\mathbf{J} = -D\nabla C$, where the proportionality

Solving Partial Differential Equation Applications with PDE2D, First Edition. Granville Sewell.
© 2018 John Wiley & Sons, Inc. Published 2018 by John Wiley & Sons, Inc.

constant $D(x, y, z)$ is the diffusion coefficient. Using Fick's law, then, our partial differential equation becomes

$$C_t = \nabla \cdot [D \nabla C] + q$$

Typical boundary conditions for this PDE may be $C = g$, that is, the density may be given, or

$$D \frac{\partial C}{\partial n} = g, \tag{3.1}$$

that is, the *inward* boundary flux $g = D \frac{\partial C}{\partial n} = D \nabla C \cdot \mathbf{n} = -\mathbf{J} \cdot \mathbf{n} = \mathbf{J} \cdot (-\mathbf{n})$ may be given ($g = 0$ on an insulated boundary). For the time-dependent problem, the initial density must also be specified.

However, if there is also convection, there is an additional flux of X that is equal to the density times the wind or current velocity vector $\mathbf{v}$ (recall that flux is density times average velocity), and then the total flux, due to both diffusion and convection, is $\mathbf{J} = -D \nabla C + C\mathbf{v}$, and the equation becomes

$$C_t = \nabla \cdot [D \nabla C - C\mathbf{v}] + q \tag{3.2}$$

Heat conduction and convection is essentially diffusion and convection of heat, that is, $C = \rho C_p T$, where ρ is the density, C_p is specific heat, and T is temperature. So the heat conduction equation is (assuming ρ, C_p are constant)

$$\rho C_p T_t = \nabla \cdot [\kappa \nabla T - \rho C_p T\mathbf{v}] + q \tag{3.3}$$

where $\kappa = D\rho C_p$ is essentially the diffusion coefficient for heat, called the conductivity.

3.2 Diffusion and Heat Conduction Examples

Example 3.1 (1D heat conduction/convection) We will solve a time-dependent heat conduction/convection problem (a 1D version of (3.3)) taken from Edsberg (2008, p. 67):

$$\rho C_p \frac{\partial T}{\partial t} + \rho C_p v \frac{\partial T}{\partial x} = \frac{\partial}{\partial x} \left(\kappa \frac{\partial T}{\partial x} \right) - k(T - T_{out}), \quad 0 \le x \le L$$

where $T(x, t)$ is the temperature of a fluid in a pipe of length $L = 5$ and $\rho = 2.5, C_p = 5.0, v = 0.0, \kappa = 3.0$ are the fluid density, specific heat, velocity, and heat conductivity coefficient, respectively. Since $v = 0$, the fluid is not moving in this model, so there is no convection. The last term on the right is a sink term $q(x)$, resulting from Newton's law of cooling, which is proportional

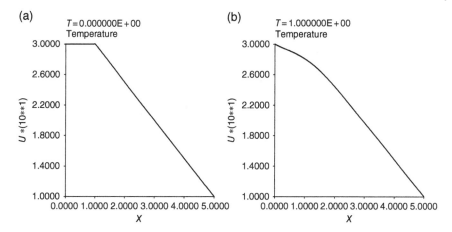

Figure 3.1 Conduction only, Example 3.1. (a) Initial temperature and (b) temperature at $t = 1$.

($k = 0.2$) to the difference between the temperature T in the pipe and the surrounding temperature $T_{out} = 10$. The initial temperature is constant at $T_0 = 30$ from $x = 0$ to $x = 1$ and then decreases linearly from T_0 to T_{out}, the temperature of the pipe wall and the outlet. The boundary conditions are $T(0, t) = T_0$ and $T(L, t) = T_{out}$. We solve this forward in time to $t = 1$ and plot the initial and final temperatures in Figure 3.1. We used the collocation method, and since "T" is reserved for time, we named the temperature "TMP."

Next we rerun with $v = 2$, so now the fluid is entering at $x = 0$ with temperature T_0 and moving from left to right, and we change the right boundary condition to a Newton's law of cooling condition:

$$\kappa T_x(L, t) = -\sigma(T(L, t) - T_{out})$$

where $\sigma = 2$. Since the right-hand side is, according to (3.1), the inward flux, this means the outward flux is proportional to the difference between the temperature at the pipe end and the outside temperature. Now the fluid entering from the left heats the pipe, and we plot the new temperature at $t = 1$ in Figure 3.2.

Finally, we repeat the problem, with $v = 2$ still, but now with no heat conduction, $\kappa = 0$. Now this is a hyperbolic transport problem, so there should now be "no" boundary condition (input as NONE) at the outlet, since it is first order in space so only one boundary condition (at the inlet) is needed. Note (Figure 3.3) that without conduction (diffusion of heat), the discontinuity in the derivative, at $x = 1$ initially, is not smoothed out now but simply transported to the right with velocity $v = 2$, so it has moved to $x = 3$ by $t = 1$.

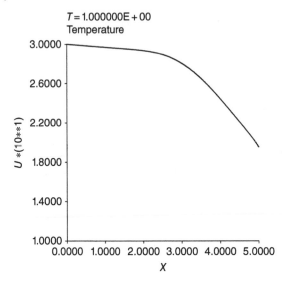

T = 1.000000E + 00
Temperature

Figure 3.2 Conduction and convection, Example 3.1.

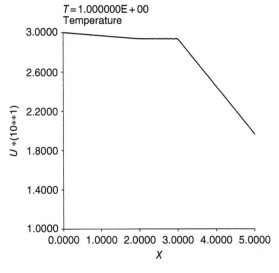

T = 1.000000E + 00
Temperature

Figure 3.3 Convection only, Example 3.1.

Figure 3.4 Diffusion Example 3.2. (a) Manually-generated initial triangulation and (b) adaptively generated final triangulation.

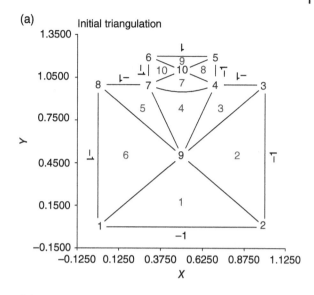

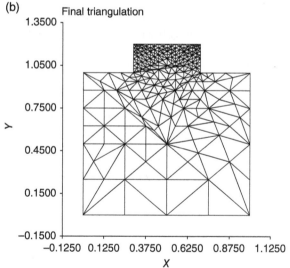

Example 3.2 (2D diffusion) As a second example, we will solve the 2D steady-state version of (3.2), without convection or sources:

$$\frac{\partial}{\partial x}[D(x,y)U_x] + \frac{\partial}{\partial y}[D(x,y)U_y] = 0,$$

in a rectangle with a "knob" on top, shown in Figure 3.4a. The diffusion coefficient $D(x,y)$ is equal to 5 in the "knob" (initial triangles 7–10) and 1 in the lower region. The interface is defined by three points (0.3,1.0), (0.5,0.95), and (0.7,1.0) through which PDE2D fits a cubic spline.

The Galerkin method is used for this 2D problem, so the fact that the diffusion coefficient is discontinuous at an interface is not a problem, even though the interface is curved. As discussed in Section I.1, $D(x, y)$ can simply be defined as a discontinuous function, and no interface conditions are needed between the two regions. In fact, we can reference the *initial* triangle number "KTRI" to define D in a Fortran block:

```
IF (KTRI <= 6) THEN
     D = 1
ELSE
     D = 5
ENDIF
```

The boundary conditions are

$$D\frac{\partial U}{\partial n} = 1 \text{ at the top of the knob.}$$

$$U = 0 \text{ on the rest of the boundary.}$$

Thus, there is a constant flux of U into (see (3.1)) the region along the top boundary, while along the rest of the boundary, the density U is held at 0.

A uniform triangulation of 500 quadratic triangles is first used, and then the problem is resolved using adaptive grid generation, resulting in the graded grid shown in Figure 3.4b. The adaptively generated grid is most dense near the top, where the solute is entering. Since the true solution will have discontinuous derivatives at the interface and the approximate solution can have discontinuous derivatives only at triangle boundaries, it is important for accuracy reasons that no triangles in the final triangulation straddle the interface. Thus, notice how the refinement of the initial triangulation follows the curved interface.

A contour plot, generated by PDE2D, of the density is shown in Figure 3.5.

Example 3.3 (3D diffusion) Next we will solve the steady-state version of the 3D diffusion problem (3.2), in the bullet-shaped region of Figure 3.6, without convection ($\mathbf{v} = \mathbf{0}$):

$$0 = \nabla \cdot [D \nabla C] + q \tag{3.4}$$

The region is a cylinder whose axis is the x-axis and whose radius varies with x: $r(x) = 1$ for $-1 \leq x \leq 0$ and $r(x) = cos(\pi x/2)$ for $0 < x \leq 1$.

We will take the diffusion coefficient $D = 5$, and the source term $q(x)$ will be 1 for $x > 0$ and zero otherwise. Thus, the diffusing element C is being created by a chemical reaction in the forward (nose cone) half of the bullet, but not in the back half. The rear surface is in contact with another material that has no C, so the boundary condition at $x = -1$ will be $C = 0$. The rest of the boundary is insulated, so the normal derivative is zero, $\frac{\partial C}{\partial n} \equiv \nabla C \cdot \mathbf{n} = 0$, that is, the flux

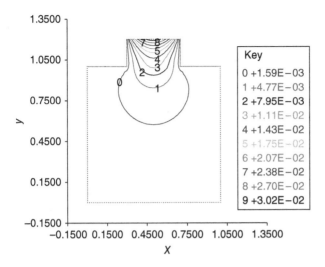

Figure 3.5 Contour plot of density, Example 3.2.

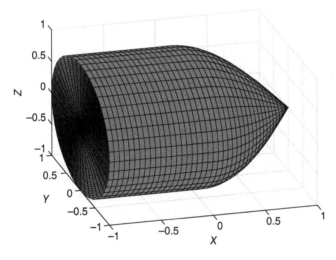

Figure 3.6 Bullet-shaped 3D region, Example 3.3.

is parallel to the boundary. For reasons of accuracy, we make sure there is a gridline at $x = 0$, where the source term $q(x)$ is discontinuous. The parametric equations supplied to define the 3D region, required by PDE2D, were [1]

---

1 The PDE2D collocation method requires that the parameterization be *smooth*: not only continuous but also continuously differentiable, and note that $r'(x)$ is continuous at $x = 0$. If $r(x) = 1 - x$, for $x > 0$, were used, for example, results might be incorrect.

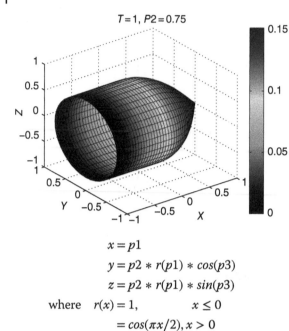

$T=1, P2=0.75$

Figure 3.7 Concentration at a cross section of bullet, Example 3.3.

$$x = p1$$
$$y = p2 * r(p1) * cos(p3)$$
$$z = p2 * r(p1) * sin(p3)$$
$$\text{where} \quad r(x) = 1, \qquad x \leq 0$$
$$= cos(\pi x/2), x > 0$$

where $-1 \leq p1 \leq 1, 0 \leq p2 \leq 1, 0 \leq p3 \leq 2\pi$. There are periodic boundary conditions at $p3 = 0, 2\pi$ (i.e. C and its derivative with respect to $p3$ are continuous) and no boundary, thus no boundary condition (enter NONE), at $p2 = 0$.

To define a coefficient, such as $r(x)$ or $q(x)$, using an IF statement, remember that you can simply reference a Fortran function and supply the function at the end (see Problem 6 of Chapter 1).

Figure 3.7 shows a MATLAB plot of the density, in one P2=constant cross section of the bullet. This plot and other cross-sectional plots are created by removing all occurrences of "C!" in the PDE2D program and running the MATLAB program "pde2d.m" automatically generated when PDE2D is executed, with the automatically generated data file "pde2d.rdm" in the same directory.

As a check on the solution, we asked PDE2D to compute the boundary flux (boundary integral of $D \frac{\partial C}{\partial n}$, that is, of "D*Cnorm") that should be (and is) equal to $-\pi/2$ because

$$\iiint\limits_{\Omega} \nabla \cdot [D \nabla C]\, dV = -\iiint\limits_{\Omega} q\, dV$$

$$\iint\limits_{\partial\Omega} D \frac{\partial C}{\partial n}\, dA = -\frac{\pi}{2}$$

That is, the outward boundary flux balances the internal generation rate.

If this 3D problem is solved in cylindrical coordinates, that is, if y, z are replaced by polar coordinates r, θ, the solution will obviously not depend on the polar angle θ, so then (3.4) can be written in the axisymmetric form (cylindrical coordinates without θ, see Problem 4d of Chapter 5):

$$0 = \frac{\partial}{\partial x}\left(D\frac{\partial C}{\partial x}\right) + \frac{1}{r}\frac{\partial}{\partial r}\left(rD\frac{\partial C}{\partial r}\right) + q \tag{3.5}$$

We resolved the 3D problem as a 2D axisymmetric problem, using the Galerkin method, and we computed the integral of C over the entire bullet (note: this means we integrate $2\pi r\, C$ over the 2D axisymmetric cross section); the resulting integral (0.3492) agreed with that calculated when the 3D problem was solved. The boundary flux (boundary integral of $2\pi r\, D\frac{\partial C}{\partial n}$) for the axisymmetric problem was again equal to $-\pi/2$. Of course we have to use "y" for "r" (PDE2D allows the user to specify the names of the solution components, but not the independent variables), and again it is important that there be no triangles in the initial triangulation (and thus none in the final triangulation) that straddle the interface $x = 0$. We have to multiply the equation through by r to put it into the divergence form required by the Galerkin method. Then on the bottom of the region, the free boundary condition is $A\, N_x + B\, N_y = r\, D\, C_x N_x + r\, D\, C_r N_y = GB$, so since $N_x = r = 0$, setting $GB = 0$ means $0 = 0$ or "no" boundary condition.

Finally, we resolved the axisymmetric problem (3.5) but now with D decreased to 1 in the "nose cone" half of the bullet ($x > 0$) only. Thus the nose cone and the back half of the bullet are made out of different materials; diffusion is slower in the nose cone. Recall that the chemical reaction that is creating C only occurs in the nose cone. It is very easy to treat composite materials such as this using the Galerkin method; one simply defines $D(x)$ to be a discontinuous function of position. Note that $D(x)$ is not constant (even though it is piecewise constant), so we cannot take it out of the brackets in $\frac{\partial}{\partial x}\left[D\frac{\partial C}{\partial x}\right]$. However, $D\frac{\partial C}{\partial x}$ must be continuous; otherwise its derivative would not exist, so $\frac{\partial C}{\partial x}$ must be discontinuous also; see the MATLAB plot of C in Figure 3.8. The collocation method cannot handle such problems as easily, because to put this term in the form required by the collocation method, we would have to write $\frac{\partial}{\partial x}\left[D\frac{\partial C}{\partial x}\right]$ as $D\frac{\partial^2 C}{\partial x^2} + \frac{\partial D}{\partial x}\frac{\partial C}{\partial x}$ and $\frac{\partial D}{\partial x}$ would be infinite at $x = 0$.

The fact that the solution has a discontinuous derivative at the interface $x = 0$ is not a problem for the Galerkin method, since its approximate solutions (Section B.3) have discontinuous first derivatives at triangle boundaries. The collocation approximate solutions, on the other hand (Section B.4), always have continuous first derivatives. The fact that the source term $q(x)$ is discontinuous in the 3D problem (3.4) did not cause difficulties for the collocation method because that only makes the *second* derivatives of the solution discontinuous.

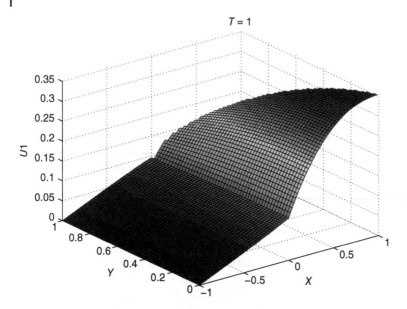

Figure 3.8 Axisymmetric solution, Example 3.3.

Example 3.4 (Image processing) An unusual application of the diffusion equation is to image processing. If a picture is converted to a matrix, where $pix(i, j)$ measures the darkness of the pixel at $(x(i), y(j))$, one could define initial conditions for a 2D diffusion problem (3.2) (with no convection or sources) by assigning $U(x, y, 0) = pix(i, j)$ when $(x, y) = (x(i), y(j))$ and interpolating values between pixels. The initial $U(x, y, 0)$ then may be very noisy, but diffusion will smooth out the solution, decreasing noise and blurring sharp interfaces, and in fact, the eventual result (assuming insulated boundary conditions, $\frac{\partial U}{\partial n} = 0$) will be $U = constant$, that is, the picture will be completely destroyed!

However, what is sometimes done (Perona and Malik 1990) to improve noisy pictures is to define a nonlinear diffusion coefficient, such as $D = 1/[1 + (U_x^2 + U_y^2)/\lambda^2]$, which is designed to be small where the gradient of U is large. In this manner, diffusion will be slow near sharp interfaces, where the gradient is large, the assumption being that sharp interfaces are real and important to the picture, while smaller gradients may be due to noise and need to be smoothed out. We still have to be careful not to let diffusion proceed too long, as the picture will eventually be destroyed no matter how slow diffusion is. But to show how this can improve a picture, Figure 3.9 shows an original picture, PDE2D read in the pixel data from this picture, did linear interpolation between pixels, and then added small random perturbations, so the initial $U(x, y, 0)$ is noisy as shown in Figure 3.10a. The final picture, after diffusion out to $t = 0.005$, is shown in Figure 3.10b. You can see that the nonlinear diffusion has smoothed out the

Figure 3.9 Original picture, Example 3.4.

noise without blurring too much the protected "real" features of the photo. To get the pictures in Figures 3.9 and 3.10, we only had to replace the call to the MATLAB surface plot routine "surf" with a call to "imagesc" in the MATLAB program generated automatically by PDE2D.

Reading the initial conditions from a file, interpolating and adding noise was relatively easy with PDE2D; we simply input the initial conditions as "UINIT(X,Y)" and then defined the Fortran function UINIT as shown below:

```
      FUNCTION UINIT(X,Y)
      IMPLICIT DOUBLE PRECISION (A-H,O-Z)
      PARAMETER (NPIX=127)
      DIMENSION PIX(0:NPIX,0:NPIX)
      SAVE PIX
      LOGICAL READ
      DATA READ /.FALSE./
      IF (.NOT.READ) THEN
C                       READ PIXEL FILE THE FIRST CALL ONLY
      OPEN (11,FILE='ex4.ch3.pix.txt')
      DO J=0,NPIX
         READ (11,*) (PIX(I,J),I=0,NPIX)
      END DO
      CLOSE (11)
      READ = .TRUE.
      ENDIF
C                       ON ALL CALLS, USE BILINEAR INTERPOLATION
C                          TO GET PIXEL VALUE AT (X,Y)
C                       FOUR NEAREST NEIGHBORS TO (X,Y) ARE
C                          (II,JJ)/NPIX,    (II+1,JJ)/NPIX,
C                          (II,JJ+1)/NPIX,  (II+1,JJ+1)/NPIX.
```

(a) $T = 0$

(b) $T = 0.005$

Figure 3.10 Image enhancement, Example 3.4. (a) Initial conditions, with noise added and (b) final picture, after nonlinear diffusion.

```
        H = 1.D0/NPIX
        II = MIN(INT(X/H),NPIX-1)
        JJ = MIN(INT(Y/H),NPIX-1)
        XX = X/H - II
        YY = Y/H - JJ
        U1 = PIX(II,JJ)   + XX*(PIX(II+1,JJ)  -PIX(II,JJ))
        U2 = PIX(II,JJ+1) + XX*(PIX(II+1,JJ+1)-PIX(II,JJ+1))
        UINIT = U1 + YY*(U2-U1)
C                           ADD RANDOM NOISE OF AMPLITUDE AMP
```

```
AMP = 30
UINIT = UINIT + AMP*(2*RAND()-1.0)
RETURN
END
```

Reading and Interpolating Initial Conditions, Example 3.4

A value of $\lambda = 200$ was used; a smaller value of λ will further slow diffusion near the protected sharp interfaces. A very fine grid is of course essential; here *pix* was a 128 by 128 matrix, and so a 128 by 128 grid was used, with 4 linear triangular elements in each grid rectangle. The Galerkin method was used because Galerkin approximations can have discontinuous derivatives, and thus noisy data can be better handled.

3.3 Problems

1 (Diffusion/reaction) Consider a diffusion–reaction problem that models the diffusion in a solid of a certain atomic element "*E*," whose density is given by v, and its diatomic molecule, E_2, whose density is u. Suppose two atoms of E bind to form a molecule with frequency proportional to v^2 (the probability that two atoms are close enough to unite is proportional to the square of the atomic density) and a molecule breaks up to form two atoms with frequency proportional to u (only one molecule is needed for a breakup). Then we have two diffusion equations of the form (3.2), with no convection:

$$u_t = D_u(u_{xx} + u_{yy}) + r_1 v^2 - r_2 u,$$
$$v_t = D_v(v_{xx} + v_{yy}) - r_1 v^2 + r_2 u$$

The reaction terms in the second equation reflect the fact that when a molecule is formed, the decrease in v must equal the increase in u, and when a molecule breaks up, the increase in v must equal the decrease in u, because two atoms weigh the same as one molecule.

Solve this diffusion–reaction problem in the unit square, with $u = 0$, $v = 1$ on the right side $x = 1$ and no-flux boundary conditions $\left(\frac{\partial u}{\partial n} = \frac{\partial v}{\partial n} = 0\right)$ on the other three sides. Take $D_u = 0.1, D_v = 2$, $r_1 = 1, r_2 = 3$, and initial conditions are $u(x, y, 0) = 0, v(x, y, 0) = 1$. You may use either Galerkin or collocation.

By $t = 4$ a steady state should be reached; make cross-sectional plots of u and v as functions of x at steady state (Figure 3.11), at any value of y (the solution will not depend on y). At steady state, there is a net flux of atoms of E rapidly diffusing in from the right boundary, binding to form E_2 molecules faster than molecules are breaking up, and the molecules are slowly diffusing out the right boundary.

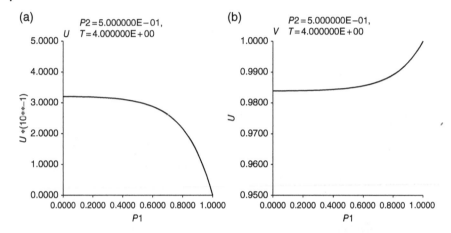

Figure 3.11 Cross sections of E_2 and E at steady state, Problem 1. (a) E_2 density vs. X and (b) E density vs. X.

Redo the problem with no-flux boundary conditions on all four sides. Now when a steady state is reached, u and v will be constant throughout the square. Predict the final values of u, v before running your program, using the facts that the rate of molecule formation now matches the rate of molecule breakup and that the total weight of atoms plus molecules of E in the region cannot change since now there is no boundary flux.

2 (2D heat convection) Solve the heat convection equation ((3.3) with no conduction or sources, $\kappa = q = 0$, $\rho\, C_p = 1$):

$$T_t = -\nabla \cdot (T\, \mathbf{v}) \tag{3.6}$$

or, if the velocity flow is incompressible, $\nabla \cdot \mathbf{v} = 0$

$$T_t = -\mathbf{v} \cdot \nabla T$$

in three-fourths of a unit circle, $0 \le r \le 1, \pi/2 \le \theta \le 2\pi$, where the incompressible velocity flow is $\mathbf{v} = (y, -x)$. The region and fluid velocity are shown in Figure 3.12. The initial condition is $T(x, y, 0) = 80$, and at the inlet boundary, $\theta = 2\pi$, you should specify $T = 100$ with no boundary conditions on the rest of the boundary. If you use the collocation method, just input NONE for these boundary conditions; if you use Galerkin, A and B should be 0, so $GB = 0$ means $0 = 0$, or no boundary condition. CRANKN=.TRUE. is not a good idea for this problem, as it will cause the solution to oscillate.

Since there is no heat conduction, the "heat front" at $\theta = 2\pi$ will simply move with the flow, and at $t = \pi/2$ one-third of the region will be "hot" ($T = 100$), while the rest will be cold ($T = 80$). If you resolve with the

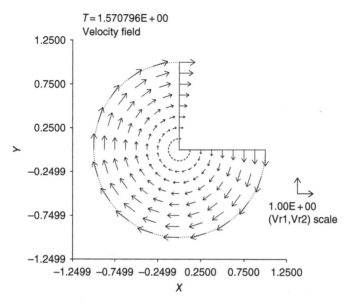

Figure 3.12 Fluid velocity field, Problem 2.

boundary condition $T = 100$ at the outlet boundary ($\theta = \pi/2$) and no boundary condition at the inlet, the solution will, not surprisingly, go unstable. Problem 1a of chapter 3 of Sewell (2015) shows that (3.6) has a unique solution if the temperature is specified on the inlet boundary only, that means, if you want to predict future temperatures, you need to know the temperatures upwind, not downwind, from where you are!

Resolve (3.3) with $\kappa = 0.01$, and you will see that heat conduction smooths out the solution (Figure 3.13b); the heat front will no longer be as sharp. Figure 3.13 shows why a small amount of "artificial diffusion" is sometimes added to a convection-only problem, to make it more tractable.

3 (Shock wave) Use either the collocation or Galerkin algorithm to solve the 1D Burgers' equation:

$$Y_t = Y \, Y_x$$
$$Y(x, 0) = x$$
$$Y(-1, t) = -1$$
$$Y(1, t) = 1$$

This nonlinear problem can be looked at as a diffusion/convection problem (3.2) with no diffusion and convection velocity $v = -Y$ dependent on the solution. As we observed in Problem 2, a first-order convection problem

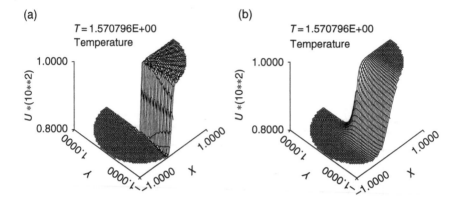

Figure 3.13 Temperature at $t = \pi/2$, Problem 2. (a) $\kappa = 0$ and (b) $\kappa = 0.01$.

Figure 3.14 Shock wave formation, Problem 3.

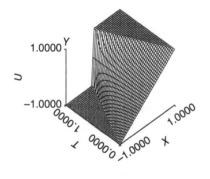

needs boundary conditions on the inlet and not the outlet. At the left end-point, $x = -1$, the velocity is held at $v = -Y = 1$, so the flow is inward; at the right endpoint $x = 1$, the velocity is $v = -Y = -1$, so the flow is inward there also, so it seems appropriate to specify boundary conditions at both ends. If you solve this problem, everything is fine until $t = 1$, and then the solution goes unstable; if you request an adaptive time step, PDE2D will take extremely small steps.

From the solution displayed in Figure 3.14, you can see what the problem is. If you think of Y as temperature, there is a flow of cold air ($Y = -1$) coming in from the left with velocity 1 and a flow of hot air ($Y = 1$) coming from the right with the same velocity, and they collide at $x = 0$ and form a discontinuous shock wave. Sometimes people encounter such behavior in nonlinear PDEs and think they just need a better numerical algorithm, but no PDE solver can get past $t = 1$ successfully on this problem, as it is the solution itself that goes unstable.

Retry the problem after adding $0.01\ Y_{xx}$ to the right-hand side of the PDE, and you will be able to integrate past $t = 1$ to a steady state. Adding

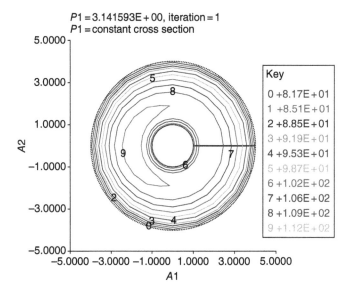

Figure 3.15 p1=π cross section, Problem 4.

a little artificial diffusion allows the hot and cold fronts to mix when they collide.

4 (Heat conduction in a torus) Solve the steady-state heat conduction problem (3.3) without convection or sources, in the hollowed out half torus of Figure I.2. That is, our torus is described by (I.1), where the toroidal angle $p1$ goes from 0 to π, the poloidal angle $p2$ goes from 0 to 2π, and $1 \leq p3 \leq 4$. Set the temperature W to 80 on the outside of the torus, $p3 = 4$, and to 100 on the inner boundary, $p3 = 1$. There is an inward heat flux ($\frac{\partial W}{\partial n} = 20$; see (3.1)) at the flat end $p1 = \pi$ and no flux ($\frac{\partial W}{\partial n} = 0$) at the other flat end, $p1 = 0$. There are periodic boundary conditions on the poloidal angle $p2$. The conductivity $\kappa = 1$ is constant, so our equation is just Laplace's equation $W_{xx} + W_{yy} + W_{zz} = 0$.

Use the GUI to create the PDE2D program, and ask for p1=constant cross sections, with axes A1=p3*cos(p2),A2=p3*sin(p2). One contour cross-sectional plot, at the flat end $p1 = \pi$ where heat is entering, is shown in Figure 3.15. Remove all occurrences of "C!" to activate production of the MATLAB graphics program, and run this MATLAB program, pde2d.m, with pde2d.rdm in the working directory, and you should get cross-sectional plots similar to Figure 3.16, which are coded by color. Note the higher temperatures in Figure 3.16b near the $p1 = \pi$ end, where heat is entering. There will also be constant poloidal angle plots.

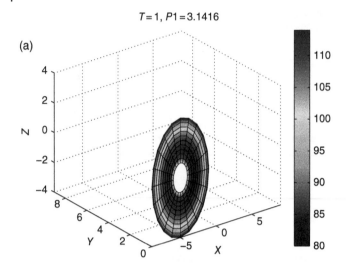

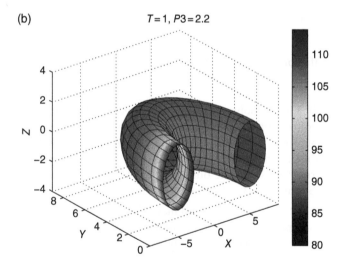

Figure 3.16 MATLAB cross-sectional plots, Problem 4. (a) $p1 = \pi$ and (b) $p3 = 2.2$.

5 (Image processing)

a) Start with initial conditions

$$U = 0.2 \quad \text{when} \quad \mod(\text{int}(x) + \text{int}(y), 2) = 0$$
$$= 0.8 \quad \text{otherwise}$$

in the square $(0, 3) \times (0, 3)$, and add some random noise of maximum amplitude 0.15 (perhaps using Fortran random number generator *rand*()), and then use nonlinear diffusion to try to smooth the noise

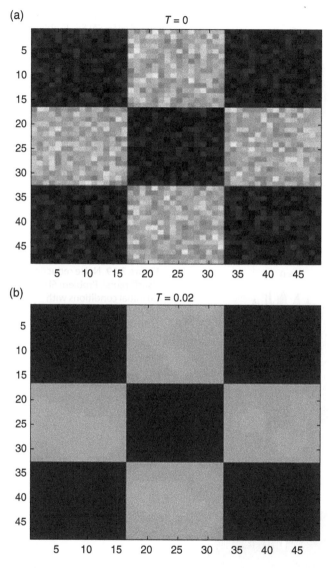

Figure 3.17 Noise reduction using nonlinear diffusion, Problem 5a. (a) Initial picture, with noise added and (b) final picture, after diffusion.

without destroying the checkerboard. Use the nonlinear diffusion coefficient $D = 1/[1 + (U_x^2 + U_y^2)/\lambda^2]$, as in Example 3.4., and solve $U_t = \nabla \cdot [D\nabla U]$ with insulating boundary conditions $\frac{\partial U}{\partial n} = 0$. Display the initial and final pictures by replacing the call to "surf" in the MATLAB program generated by PDE2D by

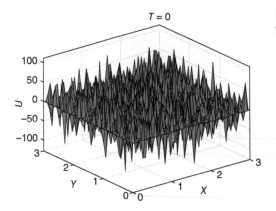

Figure 3.18 Initial conditions with LSQFIT=.FALSE., Problem 5b.

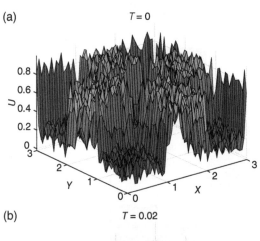

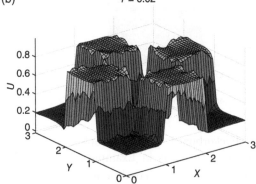

Figure 3.19 Noise reduction ("surf" plots), Problem 5b. (a) Initial conditions with LSQFIT=.TRUE. and (b) after diffusion.

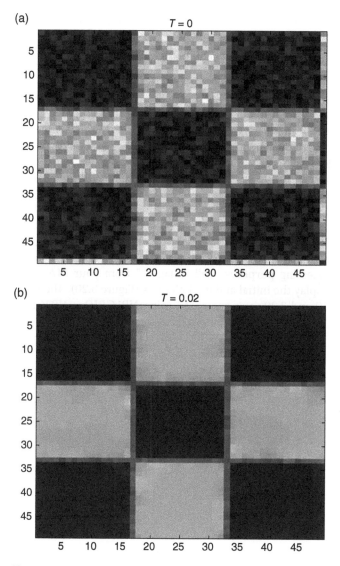

Figure 3.20 Noise reduction ("imagesc" plots), Problem 5b. (a) Initial picture, with noise added and (b) final picture, after diffusion.

```
imagesc(U(:,:,L+1,IDER,IEQ),[0 1])
colormap(gray)
```

Use the Galerkin method with IDEG=1, CRANKN=.FALSE., and ADAPT=.TRUE.. The pictures in Figure 3.17 were generated using

$NXGRID = NYGRID = 61, NX = NY = 48, \lambda = 1, t_{final} = 0.02$, but you may experiment with different parameters.

b) Repeat part (a) using the collocation method. As discussed in Section B.6, if the initial conditions have discontinuous or ill-behaved derivatives and the random noise makes these derivatives essentially infinite, you may need to use the option LSQFIT=.TRUE. when the collocation method is used. But try first with LSQFIT=.FALSE. and you will get large spikes in the initial conditions, as seen in the MATLAB plot of Figure 3.18. Notice that spikes go higher than 100, when the initial conditions should be between 0 and 1. Then switch to LSQFIT=.TRUE., and you will see much less noise (Figure 3.19a). (Remember, there is noise even in the "exact" initial conditions.)

If you had to solve the same nonlinear diffusion equation as solved by the Galerkin method in part (a), you would have to expand out $U_t = \nabla \cdot [D\nabla U] = D\nabla^2 U + \nabla D \cdot \nabla U$ for the collocation method, but there is no need to do this – the equation $U_t = D\nabla^2 U$ is good enough for image processing purposes. Call "imagesc" from your MATLAB program to display the initial and final pictures (Figure 3.20). The plots in Figures 3.19 and 3.20 were generated using $NP1GRID = NP2GRID = 61, NP1 = NP2 = 48, \lambda = 1, t_{final} = 0.02$ again.

4

Pricing Options

4.1 Derivation of Black–Scholes Equation

This section is based on an article (Sewell 2018) published in the Mathematical Association of America's *College Mathematics Journal*.

Suppose at time t your friend owns a certain asset (a stock, for example) whose market price is currently s. He hands you a signed piece of paper on which is written, "At time T, I promise to sell this asset to you for price E if you want to buy it then." How much is this piece of paper (a European "call" option) worth to you? The Black–Scholes equation (Black and Scholes 1973) is a partial differential equation that attempts to answer this question and is widely used for pricing "options" such as the one your friend offered you. This equation will be derived here from basic principles, in a way which we believe is unique in that it does not require any background in financial mathematics or stochastic calculus.

If you knew that the market price of this asset was going to be S at time T, the paper your friend offered you would be worth $P(S) \equiv max(0, S - E)$, because if S is more than E, you can make a quick profit[1] of $S - E$ by buying it from him at the "strike" price E and then reselling on the market at price S, while if S is less than E, you won't buy it, and then the option is worth nothing to you. Unfortunately, you don't know how much the asset will be worth at time T. But if you could construct a reasonable probability density function $q(S)$ for the price of the asset at T, you could calculate that the expected value of this call option is $\int_0^\infty P(S)q(S)\,dS$. Well, that is how much the option would be worth at time T, but since you are trying to decide how much it is worth right now, you have to multiply this by the factor $e^{-r(T-t)}$, where r is the inflation rate or the "risk-free interest" rate. And so, right now, the call option is worth

$$V(s,t) = e^{-r(T-t)} \int_0^\infty P(S)q(S)dS. \tag{4.1}$$

---

1 A more common scenario: You know you are going to need to buy this asset at time T, and then the value of the call option actually represents the amount you expect to save by not buying the asset at the market price.

Solving Partial Differential Equation Applications with PDE2D, First Edition. Granville Sewell.
© 2018 John Wiley & Sons, Inc. Published 2018 by John Wiley & Sons, Inc.

Suppose your friend signs a paper on which he has written, "At time T, I promise to buy this asset from you for price E if you want to sell it then." How much is this "put" option worth? Now if you knew the market price of the asset was going to be S at time T, you could say the paper would be worth $P(S) \equiv max(0, E - S)$, because if S is less than E, you could go buy it on the market for S and sell it to your friend for a profit[2] of $E - S$, while if S is more than E, you won't bother, and the option is worth nothing. Again, if you have a probability distribution for the future price, $q(S)$, the current value of the option is still given by (4.1); only the function $P(S)$ is different for a put option.

So to decide on the value of either option, you need a reasonable probability density function for the price S at time T, of an asset that is worth s at time t. A reasonable way to construct such a probability distribution, absent any insider trading information, is as follows. Let's assume that every dt years, you toss a coin (actually, the market tosses the coin); if the result is heads, the price of the asset goes up by a factor of R, which means the log of price increases by $dx \equiv ln(R)$, and if tails, the price goes down by a factor of R, that is, the log of price decreases by dx. By time T you will have tossed $N = (T - t)/dt$ coins, and the binomial distribution of the number of heads will be approximately (as is well known, assuming N is large) a normal distribution with mean of $N/2$ and standard deviation of $\sqrt{N}/2$. Thus the price S at time T will have a lognormal distribution $q(S)$, that is, the probability that the price at time T will be in a given small range dS around S is

$$q(S)dS = p(t, z)dz \equiv \frac{1}{\sqrt{2\pi\sigma^2}} exp \left[\frac{-z^2}{2\sigma^2} \right] dz \qquad (4.2)$$

where p is the normal distribution in the log of price variable $z \equiv ln(S) - \alpha$, and reasonable values for σ and α will be discussed below. (It will be shown that σ changes with time, which explains why p is a function of t as well as z.) It can be verified that $\int_0^\infty q(S)\, dS = \int_{-\infty}^\infty p(t, z)\, dz = 1$ as required of any probability distribution.

Since the most likely outcome of the coin-tossing experiment described above will be that half of the coins are heads, in which case the final price will be the same as the original price, it might seem that we should choose $\alpha = ln(s)$ so that our lognormal distribution, which has a peak at $z \equiv ln(S) - \alpha = 0$, will also have its peak at $S = s$. But the peak of a lognormal distribution is lower than its mean,[3] and it seems more reasonable that the mean, not the peak, of

---

2 A more common scenario: You already own the asset and know you are going to want to sell it at time T, and then the value of the put option represents the expected increase in income compared with selling the asset at the market price.

3 To see why, suppose the price of an asset, currently $1, doubles on heads and is cut in half on tails. After two coin tosses, there is a 25% chance the price is now $0.25, a 50% chance it is still $1, and a 25% chance it is $4. The peak of the distribution is still $1 but the mean has increased to $1.5625.

the final price distribution should be equal to the original price; in fact, taking into account inflation, it seems more reasonable that the mean should not be s, but $se^{r(T-t)}$.

Thus we will choose the shift parameter α so that the mean of the final price distribution

$$\int_0^\infty S\, q(S)\, dS = \int_{-\infty}^\infty e^{z+\alpha}\, p(t,z)\, dz = e^{\alpha+\frac{1}{2}\sigma^2}$$

will equal $se^{r(T-t)}$, which means $\alpha = ln(s) + r(T-t) - \frac{1}{2}\sigma^2$.

Now a reasonable choice for the standard deviation σ of the normal distribution of $z = ln(S) - \alpha$ is $\sigma = \frac{\sqrt{N}}{2}(2dx)$, since the standard deviation in the number of heads is $\sqrt{N}/2$ and the increase in z is (heads–tails)$dx = (2*\text{Heads-N})dx$. So the variance is $\sigma^2 = N\, dx^2 = dx^2(T-t)/dt = \sigma_1^2(T-t)$, where $\sigma_1 = dx/\sqrt{dt}$, called the "volatility," is a measure of how fast the price fluctuates.[4] Notice that σ_1 is the standard deviation in the log of price after one year. Thus

$$p(t,z) = \frac{1}{\sqrt{2\pi\sigma_1^2(T-t)}} exp\left[\frac{-z^2}{2\sigma_1^2(T-t)}\right] \tag{4.3}$$

where

$$z = ln(S) - \alpha = ln(S) - ln(s) + (\frac{1}{2}\sigma_1^2 - r)(T-t) \tag{4.4}$$

Now that we have our probability distribution, using (4.1) and (4.2) we can say that the current value of the call or put option is

$$V(s,t) = e^{-r(T-t)}\int_{-\infty}^\infty P(S)p(t,z)dz$$

or

$$V(s,t) = e^{-r(T-t)}\int_0^\infty \left[\frac{P(S)}{S}\right] p(t,z)dS. \tag{4.5}$$

where $p(t,z)$ is given by (4.3) and (4.4).

Note that the integral in (4.5) can be used to calculate the value of an option directly, without the need to solve a partial differential equation. Nevertheless, the Black–Scholes PDE can be derived from (4.3) to (4.5) as follows. First, using the product rule and the chain rule, and $\frac{\partial z}{\partial t} = r - \frac{1}{2}\sigma_1^2$, differentiating (4.5) produces

$$V_t = rV + e^{-r(T-t)}\int_0^\infty \left[\frac{P(S)}{S}\right]\left[p_t + \left(r - \frac{1}{2}\sigma_1^2\right)p_z\right]dS \tag{4.6}$$

---

4 Increasing the size dx of the variations and decreasing the time dt between variations will increase the volatility.

Similarly, using the chain rule and $\frac{\partial z}{\partial s} = -1/s$

$$V_s = e^{-r(T-t)} \int_0^\infty \left[\frac{P(S)}{S}\right] p_z \left(\frac{-1}{s}\right) dS$$

$$sV_s = -e^{-r(T-t)} \int_0^\infty \left[\frac{P(S)}{S}\right] p_z dS \qquad (4.7)$$

$$(sV_s)_s = -e^{-r(T-t)} \int_0^\infty \left[\frac{P(S)}{S}\right] p_{zz} \left(\frac{-1}{s}\right) dS$$

$$s(sV_s)_s = e^{-r(T-t)} \int_0^\infty \left[\frac{P(S)}{S}\right] p_{zz} dS \qquad (4.8)$$

so combining (4.6) and (4.8)

$$V_t + \frac{1}{2}\sigma_1^2 s(sV_s)_s = rV + e^{-r(T-t)} \int_0^\infty \left[\frac{P(S)}{S}\right] \left[p_t + \left(r - \frac{1}{2}\sigma_1^2\right)p_z + \frac{1}{2}\sigma_1^2 p_{zz}\right] dS$$

But we can verify using (4.3) that $p(t,z)$ satisfies the backward diffusion equation $p_t + \frac{1}{2}\sigma_1^2 p_{zz} = 0$ (in fact, it is a well-known[5] analytical solution to this equation), and thus

$$V_t + \frac{1}{2}\sigma_1^2 s(sV_s)_s = rV + \left(r - \frac{1}{2}\sigma_1^2\right) e^{-r(T-t)} \int_0^\infty \left[\frac{P(S)}{S}\right] p_z dS$$

and using (4.7)

$$V_t + \frac{1}{2}\sigma_1^2 s(sV_s)_s = rV - \left(r - \frac{1}{2}\sigma_1^2\right) sV_s$$

and finally, since $(sV_s)_s = sV_{ss} + V_s$

$$V_t + \frac{1}{2}\sigma_1^2 s^2 V_{ss} + \frac{1}{2}\sigma_1^2 sV_s = rV - rsV_s + \frac{1}{2}\sigma_1^2 sV_s$$

$$V_t + \frac{1}{2}\sigma_1^2 s^2 V_{ss} + rsV_s - rV = 0 \qquad (4.9)$$

Formula (4.9) is the Black–Scholes partial differential equation. Since we know how much the options are worth, as a function of asset price, at the final time T and are trying to determine the value of the option at an earlier time t, this PDE is solved backward in time, using a final condition $V(S, T) = P(S) = max(0, S - E)$ for a call option, or $V(S, T) = max(0, E - S)$ for a put option.

What are the boundary conditions for (4.9)? If the asset price s is nearly 0 at any time, it can be assumed it will still be less than E at time T, so you are sure not to want to buy the asset at the high price E, so the call option is worthless

---

5 With $-\frac{1}{2}\sigma_1^2$ replaced by D, (4.3) is the Green's function solution for $p_t = Dp_{zz}$.

to you, and $V(0, t) = 0$. The put option current value is $V(s, t) = E_c - s$, the current value of the strike price $E_c \equiv Ee^{-r(T-t)}$ minus the current price of the asset. Hence the left boundary condition can be either $V(0, t) = E_c$ or $V_s(0, t) = -1$.

If the asset price s is near some very large value, S_{max}, it can be assumed the price will still be above E at time T, and so the current value of a call option is $V(s, t) = s - E_c$, that is, the current price minus the current value of the strike price. Hence the right boundary condition can be $V(S_{max}, t) = S_{max} - E_c$ or $V_s(S_{max}, t) = 1$. For a put option, the right boundary condition will be $V(S_{max}, t) = 0$, since you are sure not to want to exercise your option to sell at the low price E.

If we can calculate the value of an option directly from the integral (4.5), why bother solving a partial differential equation? One reason is solving the Black–Scholes equation gives us the current value of the option for all possible current asset prices, not just for one value of s.

4.2 Option Pricing Examples

Example 4.1 (Black–Scholes equation) The Black–Scholes equation, for a call option, is solved on p. 117 of *Financial Engineering with Finite Elements* (Topper 2005), using $E = 40, r = 0.1, \sigma_1 = 0.2, S_{max} = 100, t = 0$, and $T = 0.5$. We solved with PDE2D using these parameters and boundary conditions $V(0, t) = 0, V_s(S_{max}, t) = 1$ and the collocation method, which is easier to set up since Black–Scholes is not in divergence format. A MATLAB plot of the

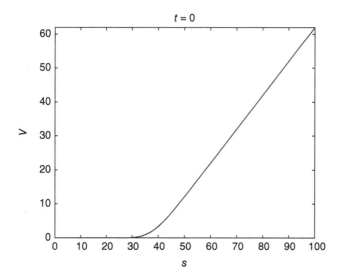

Figure 4.1 Black–Scholes solution, Example 4.1.

solution at $t = 0$ is shown in Figure 4.1. We have to use "x" to represent "s." We get a value of, for example, $V(39, 0) = 2.680$ that agrees well with Juergen Topper's analytical solution.

Note that if the current asset price is $s = 39$, the option to buy at the higher strike price, $E = 40$, is worth something (2.680), because there is some chance that it will be worth more than the strike price by time T.

This value was also confirmed using the integral (4.5):

$$V(s, 0) = e^{-rT} \int_E^\infty \left[\frac{S - E}{S} \right] p(0, z) dS$$

A midpoint rule was used to approximate this integral, with $s = 39$, in the Fortran code shown below. The result was again $V(39, 0) = 2.680$.

```
        IMPLICIT DOUBLE PRECISION (A-H,O-Z)
        PI = 3.141592654 D0
C                         NUMBER OF MIDPOINT RULE INTERVALS
        N = 10000
        Smax = 100
        E = 40
        R = 0.1
        SIGMA = 0.2
        T = 0.5
        S0 = 39
        ALPHA = LOG(S0) + (R-0.5*SIGMA**2)*T
        H = (Smax-E)/N
        SUM = 0
        DO K=1,N
C                         S AT INTERVAL MIDPOINT
        S = E + (K-0.5)*H
C                         Z FROM FORMULA 4.4
        Z = LOG(S) - ALPHA
C                         P(0,Z) FROM FORMULA 4.3
        P = EXP(-Z**2/(2*SIGMA**2*T))/SQRT(2*PI*SIGMA**2*T)
        SUM = SUM + H*(S-E)/S*P
        END DO
C                         CURRENT VALUE OF OPTION
        SUM = SUM*EXP(-R*T)
        PRINT *, ' Integral = ',SUM
        STOP
        END
```

Numerical Approximation of Integral (4.5), Example 4.1

Example 4.2 (Monte Carlo simulation) Another way to estimate the value of an option is a Monte Carlo simulation, which simulates the possible fluctuations in asset price. Start with S set to the current price s of the asset, and then every $dt = (T - t)/N$ years, flip a coin: If heads, multiply S by a factor of e^{dx}, where $dx = \sigma_1 \sqrt{dt}$ (i.e. increase $ln(S)$ by dx), and if tails, divide by this factor. After N time steps (at time T), multiply the final price by the factor $e^{r_p(T-t)}$, where $r_p = r - \frac{1}{2}\sigma_1^2$. (In this experiment, without this adjustment, the peak of the lognormal price distribution will clearly be at the starting point, $S = s$. But according to (4.4) the peak must be at $z = ln(S) - \alpha = 0$, that is, at $S = se^{r_p(T-t)}$ to ensure that the distribution has the desired mean, $se^{r(T-t)}$). Then the value at T of the option is $max(S - E, 0)$; multiply this by $e^{-r(T-t)}$ to get the current value of the option. If you repeat this entire process many times and take an average of the option values, this should estimate the true current value of the option.

We did a Monte Carlo simulation using the parameters of Example 4.1, with starting price $s = 39$, doing 10 000 000 trials with $N = 1000$ coin flips each, but distributed the trials over several processors of a multiprocessor machine, using the MPI-based Fortran program shown below. We "flip a coin" using the Fortran random number generator RAND but with a different seed on each processor; otherwise all processors would do the same simulation. This is a problem that should parallelize very well, since the only communication between processors is at the very end, where they average their averages.

With 10 000 000 trials, an average value of 2.6808 was found, in good agreement with the value of $V(39, 0)$ found using the Black–Scholes equation in Example 4.1. With one processor, the program took 308 seconds, and with 8 processors it took 41 seconds. But this Monte Carlo simulation only estimates the option value for one initial price s, while using PDE2D to solve the Black–Scholes equation produces option values for all initial prices and takes less than one second to do this, on one processor of the same machine. Furthermore, doubling the number of grid points and the number of time steps only changed $V(39, 0)$ by 0.0001 in Example 4.1, while using different seeds for the Monte Carlo random number generator changed the average option value for $s = 39$ by as much as 0.003, which suggests that the PDE2D solution may be significantly more accurate.

```
      IMPLICIT DOUBLE PRECISION (A-H,O-Z)
      INCLUDE 'mpif.h'
C                    INITIALIZE MPI
      CALL MPI_INIT (IERR)
C                    NPES = NUMBER OF PROCESSORS
      CALL MPI_COMM_SIZE (MPI_COMM_WORLD,NPES,IERR)
C                    ITASK = MY PROCESSOR NUMBER
      CALL MPI_COMM_RANK (MPI_COMM_WORLD,ITASK,IERR)
C                    TOTAL NUMBER OF TRIALS
      NTOTAL = 10000000
C                    TRIALS PER PROCESSOR
      NTRIALS = NTOTAL/NPES
```

```
        E = 40.0
        S0 = 39.0
        R = 0.1
        SIGMA = 0.2
        T = 0.5D0
        Rp = R - 0.5*SIGMA**2
C                       NUMBER OF COIN FLIPS PER TRIAL
        N = 1000
        DT = T/N
        DX = SIGMA*SQRT(DT)
        RE = EXP(DX)
        VSUM = 0.0
C                       USE DIFFERENT SEED ON EACH PROCESSOR
        CALL SRAND(ITASK)
        DO ITRIAL=1,NTRIALS
          S = S0
          DO K=1,N
            IF (RAND() > 0.5) THEN
C                       IF HEADS, MULTIPLY S BY EXP(DX)
              S = S*RE
            ELSE
C                       IF TAILS, DIVIDE S BY EXP(DX)
              S = S/RE
            ENDIF
          ENDDO
C                       ADJUST FINAL PRICE TO GET DESIRED MEAN
        S = S*EXP(Rp*T)
C                       VALUE OF OPTION AT FINAL TIME (T)
        VAL = MAX(S-E,0.D0)
        VSUM = VSUM + VAL
        ENDDO
C                       AVERAGE OVER ALL TRIALS ON PROCESSOR
        Vave = VSUM/NTRIALS
C                       CURRENT VALUE OF OPTION
        Vave = Vave*EXP(-R*T)
C                       ADD AVERAGES FROM ALL PROCESSORS
        CALL MPI_REDUCE(Vave,VaveSum,1,MPI_DOUBLE_PRECISION,
       & MPI_SUM,0,MPI_COMM_WORLD,IERR)
C                       AVERAGE OVER ALL PROCESSORS AND PRINT
        Vave = VaveSum/NPES
        IF (ITASK.EQ.0) PRINT *, ' V = ',Vave
        CALL MPI_FINALIZE(IERR)
        STOP
        END
```

Monte Carlo Simulation, Example 4.2

Example 4.3 (2D basket option) The price of a call or put option on a "basket" of two assets can be modeled by a 2D partial differential equation similar to (4.9). In Topper (2005, pp. 197–198), the value of a put option on a two-asset basket ("I promise to buy the entire basket for price E at time T, if you want to sell it then") is modeled by

$$V_t + \frac{1}{2}\sigma_1^2 x^2 V_{xx} + \frac{1}{2}\sigma_2^2 y^2 V_{yy} + \rho\sigma_1\sigma_2 xy V_{xy} + rx V_x + ry V_y - rV = 0$$

(4.10)

with final and boundary conditions

$$V(x, y, T) = max(0, E - (x + y))$$
$$V(x, 0, t) = g(x, t)$$
$$V(x, S_{max}, t) = 0$$
$$V(0, y, t) = g(y, t)$$
$$V(S_{max}, y, t) = 0$$

Here $V(x, y, t)$ is the value of a put option on a basket of assets with current (time $= t$) prices x and y, and σ_1, σ_2 are the volatilities of the two assets; r is still the "risk-free interest rate," $-1 \le \rho \le 1$ is the correlation coefficient between the two assets[6], and $g(x, t)$ is the solution of the one-asset Black–Scholes problem (4.9). With the PDE2D collocation method, you can implement the boundary conditions at $x = 0$ and $y = 0$ in a unique and simple way (Florescu et al. 2013, p. 21): You just enter a boundary condition of NONE, and PDE2D enforces the PDE itself at the boundary point instead of a boundary condition. Since when $x = 0$ or $y = 0$, the 2D equation (4.10) reduces to the Black–Scholes equation (4.9) in the other variable, this is exactly what is desired!

Topper solves (4.10) with, among other parameter values, $r = 0.1, T = 0.5, E = 40, \rho = 0.5, \sigma_1 = 0.1, \sigma_2 = 0.3, S_{max} = 100$ and reports $V(18, 20, 0) = 2.0187$. We solved the equation, using the PDE2D collocation method since we can write the PDE almost exactly as it stands, and the option value as a function of the two prices x, y at $t = 0$ is shown in the MATLAB plot of Figure 4.2. With a 100 by 100 nonuniform grid, denser in the square $0 < x < 50, 0 < y < 50$, and a time step size of 0.0025, a value of $V(18, 20, 0) = 2.0182$ was obtained from the requested tabular output.

PDE2D is used throughout (Topper 2005) to solve math finance PDEs, mostly using the collocation option since many are difficult, and some are even impossible, to put into the Galerkin divergence format. PDE2D has also been used to solve some math finance integrodifferential equations (Florescu et al. 2013; Florescu, Mariani, and Sewell 2014; Dang, Nguyen, and Sewell 2016) in which

---

6 If they are totally unrelated assets, $\rho = 0$, but ρ might be close to one if, for example, the two assets are stocks in a gold mine in Canada and a gold mine in South Africa!

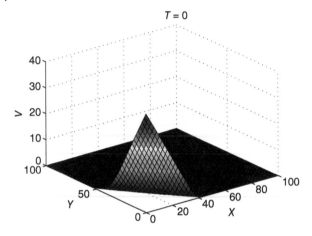

Figure 4.2 Two-asset solution of Example 4.3.

integrals of the solution appear in the PDEs. PDE2D has a feature that allows users to easily access the solution at an arbitrary point, at the previous time step or iteration, which makes it possible to handle integrals of the solution (see Problem 4).

4.3 Problems

1 (Black–Scholes equation)
  a) Solve the Black–Scholes equation (4.9), and determine the current ($t = 0$) value of a European call option that offers to sell you a stock for strike price $E = 30$ at $T = 1$; if the risk-free interest rate is $r = 0.05$, the stock volatility is $\sigma_1 = 0.1$, and the current stock price is $s = 25$. Use the collocation method, and be sure to put gridlines at and close to the strike price. (Answer: $V(25, 0) = 0.1156$) Also output the "Greeks" $V_s(25, 0)$, $V_{ss}(25, 0)$. Rerun with the volatility increased to $\sigma_1 = 0.2$, and notice that this increases the value of the option, whether the current price is more or less than the strike price. Volatility generally increases the value of any option–remember you are not buying the stock at $t = 0$, only the option to buy it later *if you want to.*
  b) Check your answer by computing the integral (4.5) directly. You can just modify the midpoint rule program used for this integral in Example 4.1. Also use the integral (4.7) to check your $V_s(25, 0)$.
  c) Resolve Problem 1a using a Monte Carlo simulation as in Example 4.2, with 1 000 000 trials, and starting price $s = 25$, so you should again get an option value of about 0.1156. There's no need to run on multiple processors, however; you can remove the MPI calls from the Example 4.2 code and set NPES=1, ITASK=0.

d) Try solving Problem 1a *forward* in time, from $t = 1$ to $t = 2$, and you will observe that the solution goes unstable quickly. The Black–Scholes equation is similar to the backward heat equation and thus can only be solved backward in time. To understand why, consider the 1D version of the diffusion or heat equation (3.2), $U_t = DU_{xx}$ with no convection or sinks or sources and with insulating boundary conditions $U_x(0, t) = U_x(L, t) = 0$. The rate of change of "entropy" in this isolated system is by the usual definition:

$$S_t = \int_0^L \frac{U_t}{U} \, dx = \int_0^L \frac{DU_{xx}}{U} \, dx = D \int_0^L \left(\frac{U_x}{U} \right)_x + \frac{U_x^2}{U^2} \, dx$$

$$= D\frac{U_x(L, t)}{U(L, t)} - D\frac{U_x(0, t)}{U(0, t)} + D \int_0^L \frac{U_x^2}{U^2} \, dx = D \int_0^L \frac{U_x^2}{U^2} \, dx$$

Thus if $D > 0$, as it is for heat conduction or diffusion, entropy increases with time, but the Black–Scholes equation is like a heat equation with $D = -\frac{1}{2}\sigma_1^2 s^2 < 0$, so entropy will decrease with time, in violation of the second law of thermodynamics. For this equation, if you want entropy to increase, you have to go backward in time!

2 (Put option) Repeat Problems 1a, 1b, and 1c, but make this a "put" option instead of a call option, and determine the value of this put option if the initial price is $s = 32$. Figure 4.3 shows the option value as a function of initial price s. For part (a) you just need to change the final conditions at $T = 1$ and the boundary conditions. For part (b) the integral (4.5) is now

$$V(s, 0) = e^{-rT} \int_0^E \left[\frac{E - S}{S} \right] p(0, z) dS$$

For part (c) you only need to change two lines from your Problem 1c code. In all cases, the option value should be about $V(32, 0) = 0.1892$. Notice that even though the current price of the asset is $s = 32$, the option to sell it at the lower price $E = 30$ is worth something, because the price might fall below E by time T.

3 (2D basket option) Resolve the two-asset problem of Example 4.3, changing only the volatilities to $\sigma_1 = 0.3$, $\sigma_2 = 0.2$ and the maturity time to $T = 0.95$. Topper reports a value $V(18, 20, 0) = 2.3555$, for comparison. One important hint: Since the problem is linear and none of the coefficients depend on t, if a constant time step is used, PDE2D is solving a large linear system each time step with the same matrix, so setting NOUPDT=.TRUE. is safe and cuts the execution time enormously when a direct solver (ISOLVE=1 or 2) is used, because the LU decomposition of the matrix computed in the first step is used on the remaining steps.

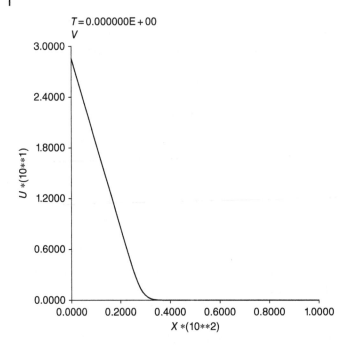

Figure 4.3 Value of put option, Problem 2a.

4 (Jump diffusion) In Florescu et al. (2013), PDE2D was used to solve a "jump diffusion" model, which is the Black–Scholes equation (4.9) with the additional terms on the right-hand side:

$$\lambda \kappa s V_s - \lambda \int_{-\infty}^{\infty} [V(se^z, t) - V(s, t)]G(z)dz$$

$$= \lambda \kappa s V_s + \lambda V - \lambda \int_{-\infty}^{\infty} V(se^z, t) \, G(z)dz \tag{4.11}$$

The additional terms are designed to model large market jumps, where $1/\lambda$ is the mean time between jumps (the time t between jumps is assumed to be a random variable with distribution $\lambda e^{-\lambda t}$) and $G(z)$ is a probability distribution for the jump factor size e^z. One of the distributions (Case B) we used in this paper was

$$G(z) = \frac{p}{\sqrt{2\pi\sigma_u^2}} e^{-\frac{(z-\mu_u)^2}{2\sigma_u^2}} + \frac{1-p}{\sqrt{2\pi\sigma_d^2}} e^{-\frac{(z-\mu_d)^2}{2\sigma_d^2}} \tag{4.12}$$

with parameters $\mu_u = 0.3753, \mu_d = -0.5503, \sigma_u = 0.18, \sigma_d = 0.6944, p = 0.3445$. Thus $G(z)$ is a mixture of two normal distributions, where there is a probability p that a sudden jump will come from an upward jump

(a)

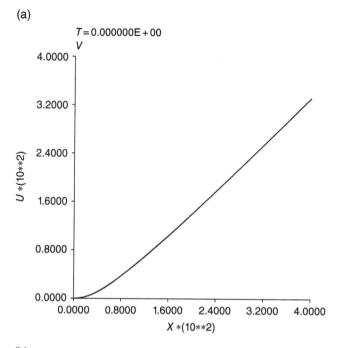

(b)

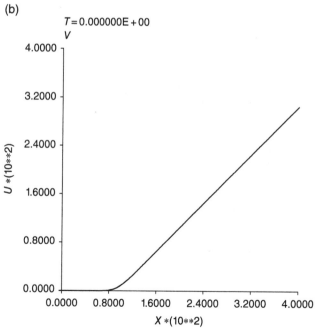

Figure 4.4 Jump diffusion model, Problem 4. (a) $\lambda = 5$ and (b) without jumps ($\lambda = 0$).

probability distribution with peak at $e^{\mu_u} = 1.4554$ (a 46% increase) and a probability $1 - p$ that it will come from a downward probability distribution with peak at $e^{\mu_d} = 0.5768$ (a 42% drop). κ is the expected value of $e^z - 1$, for this G, $\kappa = \int_{-\infty}^{\infty}(e^z - 1)G(z)dz = -0.0092591$.

Solve (4.9) with $T = 1, E = 100, \sigma_1 = 0.15, r = 0.05$, with the zero right-hand side replaced by (4.11) and $G(z)$ defined by (4.12), with a call option, that is, with a final condition of $V(S, T) = max(S - E, 0)$ and boundary conditions $V(0, t) = 0, V_s(S_{max}, t) = 1$. Take $S_{max} = 400$ and $\lambda = 5$.

Although PDE2D does not handle integral terms directly, you can approximate the integral term using any quadrature rule (e.g. a midpoint rule), with the PDE2D function $DOLDSOL1$[7] used to access the values of the solution at arbitrary points at the most recent time value. $DOLDSOL1(1, 1, X * EXP(Z), ideg)$ will interpolate (using interpolation of degree $ideg = 1, 2, 3$) the value of $V(se^z, t)$ (s is referenced as X as usual) at the last saved value of t; thus use a large value of NSAVE so that the last time value at which the solution was saved will not be too far back. NX, the number of output points saved in each time step, should also be large. Integrate for about $-4 \leq z \leq 3$, since $G(z)$ will be negligibly small outside this range. Since se^z may be well beyond S_{max}, you need to extrapolate in that case a value for $V(se^z, t) \approx V(S_{max}, t) + se^z - S_{max}$, using the fact that $V_s \approx 1$ beyond S_{max}.

In this way, PDE2D can indirectly handle integrodifferential equations, treating the integral terms in an explicit manner. Handling integrals implicitly would require the solution of a full linear system every time step and thus would not be feasible.

Print out (via tabular output or in subroutine POSTPR) values for $V(92, 0), V(100, 0)$, which should be approximately 46.20, 52.50. Figure 4.4 shows the value of this call option as a function of asset price at $t = 0$, with and without ($\lambda = 0$) jumps. Notice that the large market jumps increase the value of this call option, even though κ is slightly negative, that is, even though the expected value of the jump factor e^z is slightly less than one. For example, with $\lambda = 0, V(100, 0) = 8.6$, which means that if the current price of the asset is exactly equal to the strike price $E = 100$, adding large jumps to the model increases the value of the option from 8.6 to 52.5. The jumps make the price more volatile, and volatility generally increases the value of any option, as noted in Problem 1a.

---

7 There are also functions (D)OLDSOL2,(D)OLDSOL3 for 2D and 3D problems.

5

Elasticity

5.1 Derivation of Elasticity Equations

The equations that model displacements in an elastic body, such as a metal block, can be derived using Newton's second law. If we focus on an arbitrary small volume S of this body and set the mass times the acceleration equal to the sum of the internal and external forces acting on S, we get

$$\iiint_S \rho U_{tt}\, dx\, dy\, dz = \iint_{\partial S} (\sigma_{xx} N_x + \sigma_{xy} N_y + \sigma_{xz} N_z)\, dA$$

$$+ \iiint_S f_1\, dx\, dy\, dz$$

$$\iiint_S \rho V_{tt}\, dx\, dy\, dz = \iint_{\partial S} (\sigma_{yx} N_x + \sigma_{yy} N_y + \sigma_{yz} N_z)\, dA$$

$$+ \iiint_S f_2\, dx\, dy\, dz$$

$$\iiint_S \rho W_{tt}\, dx\, dy\, dz = \iint_{\partial S} (\sigma_{zx} N_x + \sigma_{zy} N_y + \sigma_{zz} N_z)\, dA$$

$$+ \iiint_S f_3\, dx\, dy\, dz \tag{5.1}$$

Here $\rho(x, y, z)$ is the density, and $(U(x, y, z, t), V(x, y, z, t), W(x, y, z, t))$ is the displacement vector. That is, the body element that is at (x, y, z) when the elastic body is unloaded and in equilibrium is displaced to $(x + U, y + V, z + W)$ at time t. Thus (U_{tt}, V_{tt}, W_{tt}) is the acceleration vector. The stresses σ_{pq} are the elements of a stress tensor (matrix), defined so that

$$\begin{bmatrix} \sigma_{xx} & \sigma_{xy} & \sigma_{xz} \\ \sigma_{yx} & \sigma_{yy} & \sigma_{yz} \\ \sigma_{zx} & \sigma_{zy} & \sigma_{zz} \end{bmatrix} \begin{bmatrix} N_x \\ N_y \\ N_z \end{bmatrix}$$

Solving Partial Differential Equation Applications with PDE2D, First Edition. Granville Sewell.
© 2018 John Wiley & Sons, Inc. Published 2018 by John Wiley & Sons, Inc.

gives the force per unit area that the rest of the body exerts on a point on the boundary ∂S that has unit outward normal (N_x, N_y, N_z). Thus σ_{pq} $(p, q = x, y$ or $z)$ can be interpreted as the component in the p-axis direction of the internal force experienced by a boundary element of unit area that is perpendicular to the q-axis. The vector $(f_1(x, y, z, t), f_2(x, y, z, t), f_3(x, y, z, t))$ represents the force per unit volume attributable to external sources, such as gravity.

After the divergence theorem (formula (A.1)) is applied to the boundary integrals in (5.1), we get

$$\iiint_S \rho U_{tt}\, dx\, dy\, dz = \iiint_S (\sigma_{xx})_x + (\sigma_{xy})_y + (\sigma_{xz})_z + f_1\, dx\, dy\, dz$$

$$\iiint_S \rho V_{tt}\, dx\, dy\, dz = \iiint_S (\sigma_{yx})_x + (\sigma_{yy})_y + (\sigma_{yz})_z + f_2\, dx\, dy\, dz$$

$$\iiint_S \rho W_{tt}\, dx\, dy\, dz = \iiint_S (\sigma_{zx})_x + (\sigma_{zy})_y + (\sigma_{zz})_z + f_3\, dx\, dy\, dz$$

Since S is an arbitrary volume, it can be made so small that the integrands are almost constant, so that at any point

$$\rho U_{tt} = (\sigma_{xx})_x + (\sigma_{xy})_y + (\sigma_{xz})_z + f_1$$
$$\rho V_{tt} = (\sigma_{yx})_x + (\sigma_{yy})_y + (\sigma_{yz})_z + f_2$$
$$\rho W_{tt} = (\sigma_{zx})_x + (\sigma_{zy})_y + (\sigma_{zz})_z + f_3 \tag{5.2}$$

For a linear, isotropic (direction independent) elastic body, it has been determined experimentally that

$$U_x = \frac{1}{E}[\sigma_{xx} - v\sigma_{yy} - v\sigma_{zz}]$$

$$V_y = \frac{1}{E}[-v\sigma_{xx} + \sigma_{yy} - v\sigma_{zz}]$$

$$W_z = \frac{1}{E}[-v\sigma_{xx} - v\sigma_{yy} + \sigma_{zz}]$$

$$U_y + V_x = \frac{2(1+v)}{E}\sigma_{xy}$$

$$U_z + W_x = \frac{2(1+v)}{E}\sigma_{xz}$$

$$V_z + W_y = \frac{2(1+v)}{E}\sigma_{yz} \tag{5.3}$$

where E and v are material properties called the elastic modulus and the Poisson ratio, respectively. Or solving for the stresses

$$\sigma_{xx} = E\frac{(1-v)U_x + vV_y + vW_z}{(1+v)(1-2v)}$$

$$\sigma_{yy} = E\frac{vU_x + (1-v)V_y + vW_z}{(1+v)(1-2v)}$$

$$\sigma_{zz} = E\frac{vU_x + vV_y + (1-v)W_z}{(1+v)(1-2v)}$$

$$\sigma_{xy} = \sigma_{yx} = E\frac{U_y + V_x}{2(1+v)}$$

$$\sigma_{xz} = \sigma_{zx} = E\frac{U_z + W_x}{2(1+v)}$$

$$\sigma_{yz} = \sigma_{zy} = E\frac{V_z + W_y}{2(1+v)} \tag{5.4}$$

On the boundary, normally either the displacement vector (U, V, W) is specified, or else an applied external boundary force (per unit area) vector $(g_1(x, y, z, t), g_2(x, y, z, t), g_3(x, y, z, t))$ is given. The boundary conditions that model the latter situation are obtained by balancing the internal and external forces on the boundary:

$$\begin{bmatrix} \sigma_{xx} & \sigma_{xy} & \sigma_{xz} \\ \sigma_{yx} & \sigma_{yy} & \sigma_{yz} \\ \sigma_{zx} & \sigma_{zy} & \sigma_{zz} \end{bmatrix} \begin{bmatrix} N_x \\ N_y \\ N_z \end{bmatrix} = \begin{bmatrix} g_1 \\ g_2 \\ g_3 \end{bmatrix} \tag{5.5}$$

At $t = 0$, the initial displacements (U, V, W) and the initial displacement velocities (U_t, V_t, W_t) must be specified throughout the body.

5.2 Elasticity Examples

Example 5.1 (Stresses in arch, collocation method) In a 2D plate, Eqs. (5.2) become

$$\rho U_{tt} = \frac{\partial}{\partial x}\sigma_{xx} + \frac{\partial}{\partial y}\sigma_{xy} + f_1$$

$$\rho V_{tt} = \frac{\partial}{\partial x}\sigma_{yx} + \frac{\partial}{\partial y}\sigma_{yy} + f_2 \tag{5.6}$$

For a thick plate, it is assumed that $U_z = V_z = W = 0$ and then the stress–strain relations (5.4) reduce to the "plane strain" equations:

$$\sigma_{xx} = E\frac{(1-v)U_x + vV_y}{(1+v)(1-2v)}$$

$$\sigma_{yy} = E\frac{vU_x + (1 - v)V_y}{(1 + v)(1 - 2v)}$$

$$\sigma_{xy} = \sigma_{yx} = E\frac{U_y + V_x}{2(1 + v)} \tag{5.7}$$

For a thin plate, it is assumed that $\sigma_{xz} = \sigma_{yz} = \sigma_{zz} = 0$ and then the stress–strain relations (5.3) can be resolved for the stresses to give the "plane stress" equations:

$$\sigma_{xx} = E\frac{U_x + vV_y}{1 - v^2}$$

$$\sigma_{yy} = E\frac{vU_x + V_y}{1 - v^2}$$

$$\sigma_{xy} = \sigma_{yx} = E\frac{U_y + V_x}{2(1 + v)} \tag{5.8}$$

Consider an arch (half annulus, in 2D) described in polar coordinates (using ITRANS=1) as $6 \le r \le 10, 0 \le \theta \le \pi$. We want to solve the steady-state plane strain equations (5.6) (without the time derivatives) and (5.7) in this arch. To use the collocation method, we simplify to the form

$$\frac{E}{2(1 + v)}\left(U_{xx} + U_{yy} + \frac{1}{1 - 2v}(U_{xx} + V_{yx})\right) + f_1 = 0$$

$$\frac{E}{2(1 + v)}\left(V_{xx} + V_{yy} + \frac{1}{1 - 2v}(U_{xy} + V_{yy})\right) + f_2 = 0$$

We take $E = 100$ and $v = 0.2$ and take the external force vector to be $(f_1, f_2) = (0, -10)$, that is, there is a constant downward force, namely, the weight of the uniform arch itself. On the two ends touching the "ground" ($\theta = 0, \pi$), the displacement vector is zero, $(U, V) = (0, 0)$. On the top and bottom of the arch ($r = 6, 10$), there are zero external forces, which means we must apply the boundary conditions (5.5), which for this 2D problem reduce to

$$\sigma_{xx}N_x + \sigma_{xy}N_y = g_1$$

$$\sigma_{yx}N_x + \sigma_{yy}N_y = g_2 \tag{5.9}$$

where the stresses are given by (5.7). (N_x, N_y) is the unit outward normal to the boundary, and (g_1, g_2) is the external boundary force vector, in this case $g_1 = g_2 = 0$. N_x and N_y are referenced in the boundary conditions as NORMx and NORMy. The resulting displacement vector field, (U, V), is plotted in Figure 5.1.

Example 5.2 (Stresses in arch, Galerkin method) We also solved this problem using the Galerkin method instead of collocation, using Eqs. (5.6) (again without the time derivatives) and (5.7) as written, since they are already in divergence form, and if $A1 = \sigma_{xx}, B1 = \sigma_{xy}, A2 = \sigma_{yx}, B2 = \sigma_{yy}$,

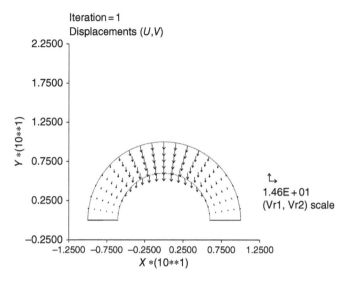

Figure 5.1 Displacement field, Example 5.1.

then (GB1,GB2) is just the boundary force vector (g_1, g_2) (see (5.9)). We used the initial triangulation option INTRI = 2, and on the free boundary, we set $(GB1,GB2) = (g_1, g_2) = (0, 0)$. The integral of V was calculated, to compare with the collocation solution, and there was good agreement (-737.4). Figure 5.2 shows the principal stresses, that is, the stresses rotated so there is no shear stress.

Example 5.3 (Stresses in 3D arch) We next solve the steady-state version of the 3D elasticity equations (5.2), with stresses defined by (5.4), in the 3D arch shown in Figure 5.3. This region was defined using ITRANS=-3 and the parametric equations:

$$X = p1$$
$$Y = p2 * cos(p3)$$
$$Z = p2 * sin(p3)$$

with $0 \leq p1 \leq 10, 6 \leq p2 \leq 10, 0 \leq p3 \leq \pi$.

This is almost a 3D version of the loaded arch problem of Examples 5.1 and 5.2, and there is again a uniform vertical external force due to the weight of the material: $(f_1, f_2, f_3) = (0, 0, -10)$. Take $E = 100, v = 0.2$ again.

The arch is fixed ($U = V = W = 0$) where it touches the ground ($z = 0$, that is, $p3 = 0, p3 = \pi$) and also at the front end $x = 0$. On the rest of the boundary, there are zero boundary forces, so Eqs. (5.5) apply with $(g_1, g_2, g_3) = (0, 0, 0)$.

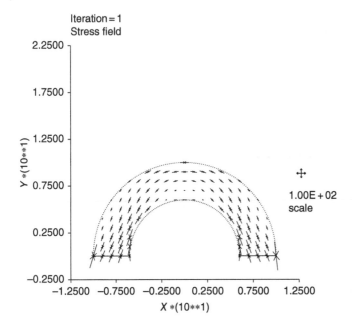

Figure 5.2 Stress field, Example 5.2.

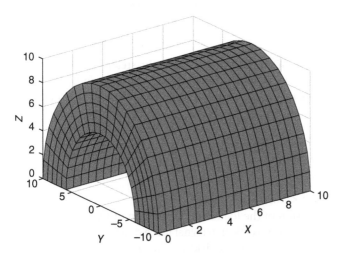

Figure 5.3 3D arch, Example 5.3.

The PDE2D coefficients $F1, F2, F3$ were defined in the form (see (5.2) and (5.4))

$$F1 = AU_{xx} + BV_{yx} + BW_{zx} + C(U_{yy} + V_{xy}) + C(U_{zz} + W_{xz})$$
$$F2 = C(U_{yx} + V_{xx}) + AV_{yy} + BU_{xy} + BW_{zy} + C(V_{zz} + W_{yz})$$
$$F3 = C(U_{zx} + W_{xx}) + C(V_{zy} + W_{yy}) + AW_{zz} + BU_{xz} + BV_{yz} - 10$$

where parameters E,VNU,A,B,C are defined:

```
E   = 100
VNU = 0.2
A   = E*(1-VNU)/(1+VNU)/(1-2*VNU)
B   =     E*VNU/(1+VNU)/(1-2*VNU)
C   = E/2.0/(1+VNU)
```

Free boundary conditions are much easier to input when a Galerkin method is used, but only collocation is available for 3D problems, so we had to write out Eqs. (5.5) and define $G1, G2, G3$ as follows:

$$G1 = (AU_x + BV_y + BW_z)N_x + C(U_y + V_x)N_y + C(U_z + W_x)N_z$$
$$G2 = C(U_y + V_x)N_x + (AV_y + BU_x + BW_z)N_y + C(V_z + W_y)N_z$$
$$G3 = C(U_z + W_x)N_x + C(V_z + W_y)N_y + (AW_z + BU_x + BV_y)N_z$$

Figures B.2 and B.3 show how these PDEs and boundary conditions can be input to the GUI almost exactly as written above (but with N_x, N_y, N_z represented by NORMx,NORMy,NORMz).

We plotted the displacements at cross sections $x = 5, 10$ (recall that $U = V = W = 0$ at $x = 0$) and chose ITPLOT=1 so that vector plots of (V, W) were made with the out-of-plane component U shown in a superimposed contour plot (Figure 5.4). The scaling is set the same in both cross-sectional plots, so it is easier to compare them.

5.3 Problems

1 (Problem with rolling friction boundary) Consider the block shown in Figure 5.5, with corners at $(0,0), (1,0), (2,1), (0,1)$. We apply a horizontal force $(GB1, GB2) = (1,0)$ to the left edge, no forces along the top, and the bottom is glued to the floor ($U = V = 0$). The right edge is flush with a 45° inclined hill, where we have a "rolling friction" or "free-slip" boundary condition. For the general 3D problem, rolling friction means the displacement component normal to the boundary is zero, $UN_x + VN_y + WN_z = 0$, and the boundary force vector (g_1, g_2, g_3) (see (5.5)) is parallel to (N_x, N_y, N_z), so that there is no boundary force component tangential to the boundary. (But see Problem 1 of Chapter 6 for an equivalent and simpler formulation of the rolling friction boundary condition.)

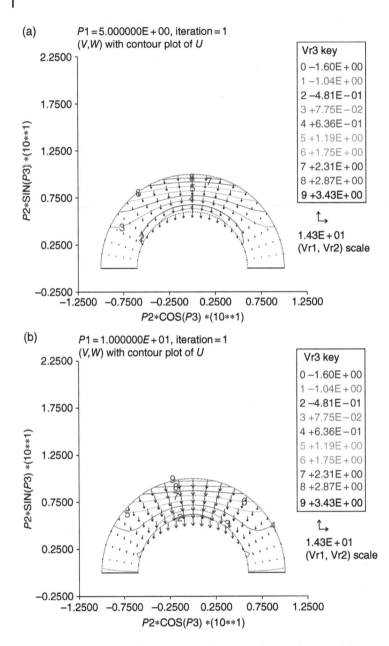

Figure 5.4 Displacement field cross sections, Example 5.3. (a) X=5 and (b) X=10.

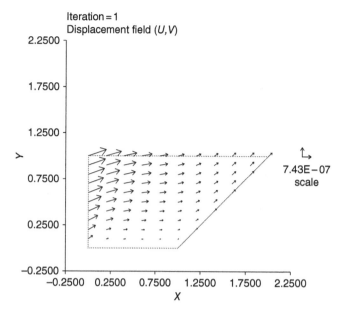

Figure 5.5 Displacement field, Problem 1.

For this 2D problem, using the Galerkin method, the rolling friction boundary condition can be enforced on a boundary with arbitrary unit normal vector (N_x, N_y) by setting

$$GB1 = \beta(UN_x + VN_y)N_x$$
$$GB2 = \beta(UN_x + VN_y)N_y$$

where β is a large number. This ensures that $UN_x + VN_y \approx 0$ and that the boundary force vector $(GB1, GB2)$ will be parallel to the normal. If β is large and negative, this can be interpreted as follows: If there is a small displacement component $UN_x + VN_y$ in the direction of the outward normal, the wall resists with a large force in the opposite direction, $(-N_x, -N_y)$. But this boundary condition actually works just as well when β is large and positive.

If $E = 10^6$, $v = 0.2$, and there are no external forces ($f_1 = f_2 = 0$), solve this steady-state 2D elasticity problem using Eq. (5.6) (without the time derivatives) and the plane stress relations (5.8), and plot the displacement field (U, V) and the stress field. Recall that the normal vector is represented by (NORMx,NORMy) in the PDE2D input.

Note that on a horizontal boundary, the rolling friction condition would reduce to $GB1 = 0$, $GB2 = \beta V$ while on a vertical boundary, $GB1 = \beta U$, $GB2 = 0$.

2 (Stresses at corner) Suppose the L-shaped block shown in Figure 5.6 (there are edges at $x = 0, 1, 2$ and $y = 0, 1, 2$) is attached ($U = V = 0$) at the base ($y = 0$), and a unit boundary force to the left ($GB1 = -1, GB2 = 0$) is applied along the top ($y = 2$), and there are zero boundary forces on the other edges. Solve (use Galerkin) the 2D elasticity problem (5.6) (without the time derivatives) with plane strain relations (5.7), with $E = 10^6$, $v = 0.1$. Plot the displacement (U, V) field and the stress field. Since stresses will be high near the corner $(1, 1)$, you should define TRIDEN to be large near this point, so the final triangulation will be dense there, as in Figure 5.6a.

3 (Axisymmetric elasticity problem) The axisymmetric equations, derived in Problem 4, are another way to reduce the 3D elasticity equations (5.2)/(5.4) to two dimensions. This means the 3D region and the forces and boundary conditions are such that if we convert the equations to cylindrical coordinates r, θ, z, the solution will not depend on the angle θ, only on r and z:

$$\rho U_{tt} = (\sigma_{rr})_r + (\sigma_{rz})_z + \frac{E(U_r - U/r)}{(1 + v)r} + f_r$$

$$\rho W_{tt} = (\sigma_{zr})_r + (\sigma_{zz})_z + \frac{E(U_z + W_r)}{2(1 + v)r} + f_z \tag{5.10}$$

with

$$\sigma_{rr} = E\frac{(1 - v)U_r + v(W_z + U/r)}{(1 + v)(1 - 2v)}$$

$$\sigma_{zz} = E\frac{v(U_r + U/r) + (1 - v)W_z}{(1 + v)(1 - 2v)}$$

$$\sigma_{rz} = \sigma_{zr} = E\frac{U_z + W_r}{2(1 + v)} \tag{5.11}$$

Here U and W are the displacements in the r and z directions, and f_r, f_z are the external forces in the r and z directions. On free boundaries, the boundary conditions have the form

$$\sigma_{rr}N_r + \sigma_{rz}N_z = g_r$$

$$\sigma_{zr}N_r + \sigma_{zz}N_z = g_z \tag{5.12}$$

where (g_r, g_z) is the boundary force vector.

Solve the steady-state axisymmetric equations in the notched cylinder of height 2 and radius 2, shown in Figure 5.7 (the vertices of the notch are $(2, 0.9), (1.5, 1.0), (2, 1.1)$). Rotate this figure with the left edge as the axis to imagine the original 3D figure. The cylinder is glued to the floor at the bottom ($U = W = 0$), and there is an upward force at the top: $(g_r, g_z) = (0, 1)$. At $r = 0$ we have symmetry boundary conditions, $U = W_r = 0$ (see Problem 1e of Chapter 6), and thus also $U_z = 0$ and $\sigma_{rz} = 0$. See Problem 2 of Chapter 2 for help on how to handle mixed boundary conditions like these using

(a)

Final triangulation

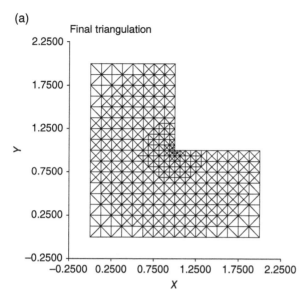

(b)

Iteration = 1
Displacement field

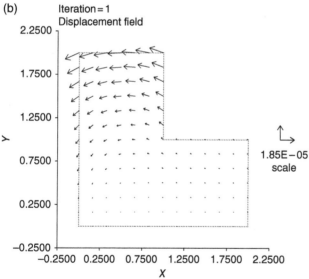

Figure 5.6 L-shaped block, Problem 2. (a) Graded final triangulation and (b) displacement field.

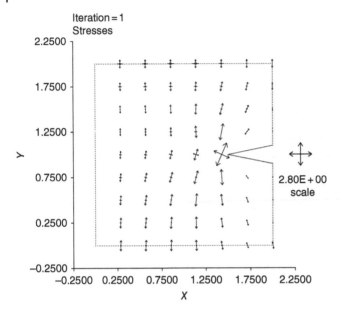

Figure 5.7 Stress field, axisymmetric Problem 3.

the Galerkin method. On the rest of the boundary, there are zero boundary forces.

Take $E = 10^6$, $v = 0.1$, $f_r = f_z = 0$, and you will have to rename r, z to x, y. Use the Galerkin method and request an adaptive triangulation, which means you need to run the problem twice, the second time the triangulation will be refined adaptively, based on the solution saved the first run. Hopefully the adaptive triangulation will be dense near the crack tip, where the stresses are large. Plot the displacement and stress fields.

4 (Derivation of axisymmetric elasticity equations)
a) Derive the first of the two axisymmetric elasticity equations (5.10) and (5.11) given in Problem 3 from the 3D equations (5.2)/(5.4). Start with the fact that $U = R\cos(\theta)$, $V = R\sin(\theta)$, where R is the displacement in the r direction. Recall that $R(r, z)$ and $W(r, z)$ are functions of r and z only, and use the chain rules:

$$F_x = F_r r_x + F_\theta \theta_x = F_r \cos(\theta) - F_\theta \sin(\theta)/r$$
$$F_y = F_r r_y + F_\theta \theta_y = F_r \sin(\theta) + F_\theta \cos(\theta)/r$$

to show that

$$U_x = R_r \cos^2(\theta) + R\sin^2(\theta)/r$$
$$U_y = R_r \sin(\theta)\cos(\theta) - R\sin(\theta)\cos(\theta)/r$$

$$U_z = R_z cos(\theta)$$
$$V_x = R_r sin(\theta)cos(\theta) - Rsin(\theta)cos(\theta)/r$$
$$V_y = R_r sin^2(\theta) + Rcos^2(\theta)/r$$
$$V_z = R_z sin(\theta)$$
$$W_x = W_r cos(\theta)$$
$$W_y = W_r sin(\theta)$$
$$W_z = W_z$$

Then you need to apply the chain rules given above to each of the x and y derivatives in the first two equations of (5.2) $((\sigma_{xx})_x$, for example); this will be an *extremely* tedious calculation! Finally, multiply the first equation by $cos(\theta)$ and the second by $sin(\theta)$ and add these two equations to get (after *much* simplification) the first equation of (5.10). Hint: To simplify the algebra a little, you may want to write Eqs. (5.4) in terms of constants $A = E(1-v)/[(1+v)(1-2v)], B = Ev/[(1+v)(1-2v)], C = E/[2(1+v)]$, and use $A = B + 2C$. You should also use $f_1 = f_r cos(\theta), f_2 = f_r sin(\theta)$, and of course R is renamed U in (5.10) and (5.11).

b) Derive the second of the two axisymmetric equations (5.10) and (5.11) from the third equation of (5.2). This also involves replacing the x and y derivatives using the chain rule and simplifying, as outlined in part (a). However, the derivation of the first equation is so much more time consuming that the instructor may want to only assign part (b) of this problem and give extra credit for solving (a)!

c) Derive the axisymmetric free boundary conditions (5.12) (with stresses given in (5.11)) from the 3D free boundary conditions (5.5) (with stresses given in (5.4)). Hints: Replace $U_x, U_y, ...$ using the formulas in part (a), and use $N_x = N_r cos(\theta), N_y = N_r sin(\theta), g_1 = g_r cos(\theta), g_2 = g_r sin(\theta), g_3 = g_z$. Then multiply the first equation of (5.5) by $cos(\theta)$ and the second by $sin(\theta)$ and add the two equations and simplify to get the first equation of (5.12). This will be much easier than deriving the first equation of (5.10) (part (a)) because it only involves first derivatives. Deriving the second equation of (5.12) will be even easier.

d) Let us also derive the axisymmetric equivalent of $(DU_x)_x + (DU_y)_y + (DU_z)_z$, first seen in Example 3.3, where $D(r,z)$ and $U(r,z)$ are independent of θ. Using the chain rule as given in part (a), show that

$$(DU_x)_x = [DU_r cos(\theta)]_x$$
$$= [DU_r cos(\theta)]_r cos(\theta) - [DU_r cos(\theta)]_\theta sin(\theta)/r$$
$$(DU_y)_y = [DU_r sin(\theta)]_y$$
$$= [DU_r sin(\theta)]_r sin(\theta) + [DU_r sin(\theta)]_\theta cos(\theta)/r$$

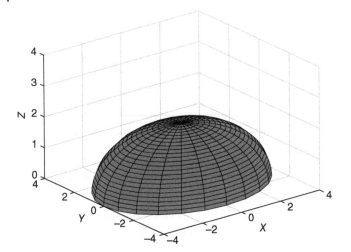

Figure 5.8 Half ellipsoid, Problem 5.

and then expand to give

$$[DU_x]_x + [DU_y]_y + [DU_z]_z = \frac{1}{r}[rDU_r]_r + [DU_z]_z$$

5 (Stresses in ellipsoid) Solve the steady-state 3D elasticity equations (5.2)/
(5.4) in the top half of an ellipsoid $(\frac{x}{4})^2 + (\frac{y}{3})^2 + (\frac{z}{2})^2 \leq 1$ shown in Figure 5.8.
There are no external forces, and take $E = 100$, $v = 0.2$, and apply a down-
ward boundary force (see (5.5)) $(g_1, g_2, g_3) = (0, 0, -1)$ on the curved top and
rolling friction conditions on the bottom.
You can define the half ellipsoid using parametric equations:

$$X = 4 * p1 * sin(p2) * cos(p3)$$
$$Y = 3 * p1 * sin(p2) * sin(p3)$$
$$Z = 2 * p1 * cos(p2)$$

with $0 \leq p1 \leq 1, 0 \leq p2 \leq \pi/2, 0 \leq p3 \leq 2\pi$ ($p1, p2, p3$ are similar to spherical coordinates ρ, Φ, θ). See Example 5.3 for how to define the PDEs and
the downward boundary force along the curved top ($p1 = 1$). On the bottom ($p2 = \pi/2$), the rolling friction boundary condition means no vertical displacements ($W = 0$) and no horizontal forces ($GB1 = GB2 = 0$, so
$\sigma_{xz} = \sigma_{yz} = 0$ since the unit normal vector is $(0, 0, -1)$). There are periodic
boundary conditions at $p3 = 0, 2\pi$ and no boundary conditions (NONE) at
$p2 = 0$, since there is no boundary there. With these boundary conditions,
however, U and V are still not uniquely determined, as only their derivatives
appear in the PDEs and boundary conditions; so also set $U = V = W = 0$
at the origin, that is, at the center point of the bottom face, $p1 = 0$.

Make a plot of the displacements (U, V) on the bottom as in Figure 5.9a (W is zero there).

6 (Elastic resonance) If we look for time periodic solutions to (5.6) with zero external forces, of the form $(U(x, y, t), V(x, y, t)) = (P(x, y)\sin(\omega t), Q(x, y) \sin(\omega t))$, we find that P, Q, ω must satisfy the eigenvalue equations:

$$-\rho\omega^2 P(x, y) = \frac{\partial}{\partial x}\sigma_{xx} + \frac{\partial}{\partial y}\sigma_{xy}$$

$$-\rho\omega^2 Q(x, y) = \frac{\partial}{\partial x}\sigma_{yx} + \frac{\partial}{\partial y}\sigma_{yy}$$

where the stresses are given by (5.7) (plane strain) or (5.8) (plane stress), with U, V replaced by P, Q.

With homogeneous boundary conditions, for most values of ω the only solution will be $P = Q = 0$ everywhere, but for certain values (the eigenvalues) there will be nonzero solutions. Find the three smallest eigenvalues of this eigenvalue problem, and plot the corresponding vector eigenfunction (P, Q) for each, in the L-shaped block of Problem 2, if $P = Q = 0$ on the bottom and there are zero forces on the rest of the boundary. Use $\rho = 0.5$ and the same values of E, ν as in Problem 2 and again use the plane strain relations (5.7). To find the smallest eigenvalue, you can set EV0R = 0, but to find the next two you may want to first set ITYPE=4 and find all eigenvalues[1] (without eigenfunctions) to get an idea of where to "go fishing" for the other two. (The smallest will be $\omega \approx 480$.) Figure 5.10 shows the vector eigenfunction corresponding to the third smallest eigenvalue; since eigenvectors/eigenfunctions are not unique, PDE2D normalizes the eigenfunction to have largest component equal to one.

Notice that if ω^2 is an eigenvalue, ω will be a natural resonant frequency for the block, because you have found nonzero solutions to (5.6) that vibrate with frequency ω, when there are no external or boundary forces.

7 (User-supplied linear system solvers) Since most of the computer time used for 2D and (especially) 3D problems is spent solving large linear systems, the speed of the linear system solver used is critical for overall efficiency. Thus PDE2D makes it easy to "plug in" other linear system solvers and test them on a wide range of large and interesting symmetric and nonsymmetric linear systems.

---

1 All eigenvalues of the discrete problem, that is. There are an infinite number of eigenvalues of the PDE problem, but the discrete eigenvalues only approximate the smaller ones.

(a)

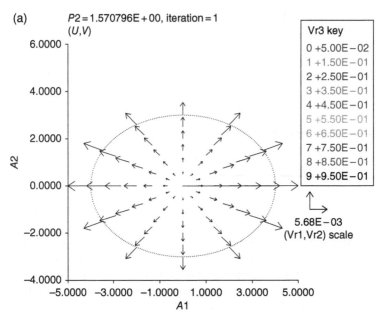

(b)

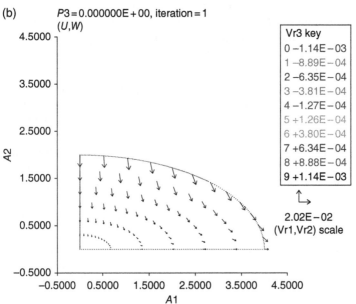

Figure 5.9 Displacement field cross sections, Problem 5. (a) Z = 0 and (b) Y = 0.

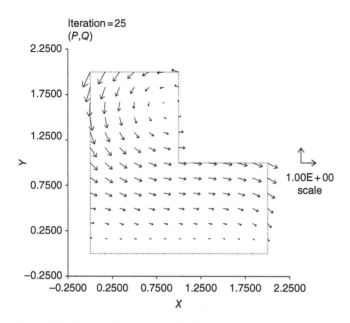

Figure 5.10 Vector eigenfunction for third eigenvalue, Problem 6.

a) Download the program ex2.ch5.f, which uses the Galerkin method to solve the 2D elasticity Example 5.2 of this chapter, from the PDE2D website (www.pde2d.com), where all examples from this book can be downloaded. Also get the file "problem57.f," which contains the plug-in routines DTD3M (for symmetric systems) and DTD3N (for nonsymmetric systems), modified to call the Jacobi conjugate-gradient iterative solver DCG. If you do not have the PDE2D source code (which is available but not free with the book), you can request (from the "Free with Book" page at www.pde2d.com) a copy of the PDE2D library with DTD3M,DTD3N removed, so that when you supply your modified versions you will not get a linker error. DCG, a slightly modified version of figure 1.10.1 of Sewell (2014), is also included in problem57.f and listed, along with the copy of DTD3M that calls it, in Section D.1.

Append the file problem57.f to ex2.ch5.f (or just add *include 'problem57.f'* to the end of ex2.ch5.f), and modify "runpde2d" to link to the special library with DTD3M, DTD3N removed. Then set ISOLVE = 5 in ex2.ch5.f, and run this program, to solve the symmetric 2D elasticity Example 5.2, with DCG used to solve the linear system. Notice that only the upper triangular nonzeros are passed to symmetric solver DTD3M; when the last argument of DCG is SYMM=.TRUE., DCG knows that the matrix is symmetric and that A will only contain the nonzeros in the upper triangle. Verify that you are getting the same results as with ISOLVE = 4.

b) Next modify your program from part (a) and set $\sigma_{yx} = 1.2\sigma_{xy}$, so the system (5.6) is now nonsymmetric. If you run with SYMM=.TRUE. still you should get a warning message that the problem is not really symmetric. So reset SYMM=.FALSE. and rerun (with ISOLVE = 5 still) and now PDE2D will call the modified version of the nonsymmetric solver DTD3N, listed in Section D.1, instead of DTD3M, and pass in all the nonzeros. DTD3N again calls DCG, but this time with SYMM=.FALSE., so DCG will know that A contains all the nonzeros. DCG is primarily designed for symmetric systems, but it can solve some moderately non-symmetric problems like this one. Finally, rerun with ISOLVE = 4 and compare the integral output to verify that DCG (ISOLVE = 5) is giving the correct solution.

Notice that DTD3M and DTD3N are called with JOB=3 when the matrix is unchanged since the last call. Thus if a direct local solver were used, it could compute and save the LU decomposition when JOB=2 and use this to solve the system much more rapidly when JOB=3.

c) Download the program ex3.ch5.f, which uses the collocation method to solve the 3D elasticity Example 5.3, and append the file problem57.f to it also. If you set ISOLVE=4, PDE2D will call DTD3M with the "normal equations" ($A^TAx = A^Tb$; see Section B.4), and the modified version of DTD3M will again call DCG to solve this symmetric positive-definite system. The matrix A^TA is denser than the original collocation matrix A, but it is still sparse; in fact it has the same nonzero structure as a Galerkin method would have for this problem. Verify that results are the same as with ISOLVE = 1.

You can also try ISOLVE=5 and then PDE2D will pass the original nonsymmetric system $Ax = b$ to DTD3N, which will try to solve it using DCG. But there is no chance DCG will be successful; in fact, the original matrix has some zeros along the diagonal, so the diagonal precondi-tioning will result in a zero divide. Other iterative methods are equally unsuccessful (maybe you can design one that will work!), because the col-location matrices are not only highly nonsymmetric, but their nonzero structures are not even symmetric. Even sparse direct solvers have diffi-culty solving these collocation systems efficiently. Even though the nor-mal equations are more ill conditioned than the original nonsymmetric system, which tends to slow convergence of iterative methods, sparse direct (ISOLVE=1, for example) and iterative (ISOLVE=3, for example) solvers generally perform *much* better on the normal equations.

d) (*Requires PDE2D source code.) If you have access to a parallel system with MPI software, and if you have the PDE2D source code, you can remove all occurrences of "C#" in the files dsubs.f, subs.f, and mach.f, to activate the MPI calls, and recompile, and link to your MPI library, and then you can run PDE2D using multiple processors. In this case, you can also test your own parallel linear system solver, and compare with PDE2D's parallel options.

As documented in the comments of DTD3M and DTD3N, if you run on more than one processor, PDE2D will distribute the columns of the large matrix cyclically over the processors (as in Figure B.4), and when it calls DTD3M or DTD3N, it will only pass the nonzeros of A corresponding to that processor's columns. (It will pass the entire right-hand side b vector to all processors.) Download the file "problem57d.f," and append it to ex3.ch5.f, or add *include 'problem57d.f'* to the end of ex3.ch5.f. This file has a copy of DTD3M, which calls PCG to solve a distributed symmetric system, and a copy of DTD3N, which calls PBAND to solve a distributed nonsymmetric system. PCG is a parallelized version of DCG and is a slightly modified version of the parallel Jacobi conjugate-gradient code shown in figure 6.2.4 of Sewell (2014). PBAND is a parallelized band solver; see Section B.8 for the details. PCG and PBAND, and the copies of DTD3M,DTD3N that call them, are displayed in Section D.2.

Table 5.1 3D local system solvers.

Unknowns = 34 560, half-bandwidth= 4541			
ISOLVE	NPES	CPU time (s)	Per-processor memory (Mwords)
1	1	463	139.2
2	1	3337	18.8
3	1	1037	20.3
3	4	306	10.3
3	16	198	2.8
→4	1	1393	54.0
→4	4	400	27.2
→4	16	254	7.1
→5	1	2349	490.6
→5	4	711	127.7
→5	16	354	32.0
6	1	1517	490.9
6	4	389	127.8
6	16	333	32.1

Now rerun ex3.ch5.f with ISOLVE=4 and PDE2D will call PCG to solve the symmetric normal equations. Also run with ISOLVE=5 and PBAND will be called to solve the very nonsymmetric original equations. PCG has no chance of solving the original equations, for the same reason as DCG, but a band solver is not as affected by the highly nonsymmetric structure of the collocation matrices. Run each test with 1 and 4 processors.

Table 5.1 shows the performance of PCG (ISOLVE=4) and PBAND (ISOLVE=5) on the problem (ex3.ch5.f) of Table I.4, on the same UTEP multiprocessor machine, with the results for the other PDE2D solvers repeated for comparison. Not surprisingly, PCG is similar in speed to the iterative solver ISOLVE=3, and PBAND is similar to the parallel band solver ISOLVE=6, though both are somewhat slower than the built-in solvers.

Although these linear system solvers are not state of the art, they are useful to illustrate the mechanics of plugging in local solvers, and now you can solve any of the other examples or problems in this book using your favorite linear systems code. Thus PDE2D provides an easy way to generate a wide range of large symmetric and nonsymmetric test problems to compare direct or iterative linear system solvers.

6

Incompressible Fluid Flow

6.1   Derivation of Navier–Stokes Equations

The Navier–Stokes equations for the flow of an incompressible fluid (a liquid) are

$$\rho(U_t + U\,U_x + V\,U_y + W\,U_z) = f_1 - P_x + \mu(U_{xx} + U_{yy} + U_{zz})$$
$$\rho(V_t + U\,V_x + V\,V_y + W\,V_z) = f_2 - P_y + \mu(V_{xx} + V_{yy} + V_{zz})$$
$$\rho(W_t + U\,W_x + V\,W_y + W\,W_z) = f_3 - P_z + \mu(W_{xx} + W_{yy} + W_{zz})$$
$$U_x + V_y + W_z = 0 \tag{6.1}$$

where (U, V, W) is the fluid velocity vector and ρ, μ, P are the fluid density, viscosity, and pressure, respectively and (f_1, f_2, f_3) is an external force vector.

Let us first derive the last equation, $U_x + V_y + W_z = 0$, called the divergence equation.

In Section 3.1 we derived the diffusion/convection/reaction equation (3.2):

$$\rho_t = \nabla \cdot [D\nabla\rho - \rho\mathbf{v}] + q$$

where $\rho(x, y, z, t)$ (called C in Section 3.1) is the density of a diffusing substance and $\mathbf{v} = (U, V, W)$ is the velocity of the medium. If there is no diffusion ($D = 0$) and no reaction (source or sink) terms ($q = 0$), so only convection is operative, we get the "continuity" equation:

$$\rho_t + \nabla \cdot [\rho\mathbf{v}] = 0$$

or

$$\rho_t + (\rho U)_x + (\rho V)_y + (\rho W)_z = 0$$

$$\rho_t + \rho_x U + \rho_y V + \rho_z W + \rho(U_x + V_y + W_z) = 0 \tag{6.2}$$

If we follow a given small piece of fluid with position $(X(t), Y(t), Z(t))$ as it moves with the fluid and define the density of this moving piece as $\rho_0(t) \equiv$

Solving Partial Differential Equation Applications with PDE2D, First Edition. Granville Sewell.
© 2018 John Wiley & Sons, Inc. Published 2018 by John Wiley & Sons, Inc.

$\rho(X(t), Y(t), Z(t), t)$, using the chain rule and the continuity equation (6.2) and remembering that $(X', Y', Z') = (U, V, W)$, we get

$$\rho_0'(t) = \rho_t + \rho_x U + \rho_y V + \rho_z W = -\rho_0(t)(U_x + V_y + W_z)$$

This means that an incompressible fluid ($\rho_0'(t) = 0$) satisfies the divergence equation $U_x + V_y + W_z = 0$. It also means that if a fluid is almost incompressible, so that it requires a large pressure P to make a small relative change in density, $\frac{P}{\alpha} = \frac{\rho_0'}{\rho_0}$, where α is a large number, then $P = -\alpha(U_x + V_y + W_z)$, which is the assumption that will be used later in the penalty formulation of the Navier–Stokes equations.

The first three ("momentum") equations in (6.1) are derived, like the elasticity equations (5.2), from Newton's second law. But now we must apply this law to a *moving* piece of fluid, which has position $(X(t), Y(t), Z(t))$, and velocity $(U(X(t), Y(t), Z(t), t), V(X(t), Y(t), Z(t), t), W(X(t), Y(t), Z(t), t))$ and density $\rho_0(t) = \rho(X(t), Y(t), Z(t), t)$. Since, as discussed above, the density $\rho_0(t)$ of a moving piece of an incompressible fluid is actually constant, setting mass (per unit volume) times acceleration (derivative of velocity) of the fluid piece equal to the force acting on the piece gives

$$\rho_0(t)\frac{d}{dt}U(X(t), Y(t), Z(t), t) = (\sigma_{xx})_x + (\sigma_{xy})_y + (\sigma_{xz})_z + f_1$$

$$\rho_0(t)\frac{d}{dt}V(X(t), Y(t), Z(t), t) = (\sigma_{yx})_x + (\sigma_{yy})_y + (\sigma_{yz})_z + f_2$$

$$\rho_0(t)\frac{d}{dt}W(X(t), Y(t), Z(t), t) = (\sigma_{zx})_x + (\sigma_{zy})_y + (\sigma_{zz})_z + f_3$$

or using the chain rule and $(X', Y', Z') = (U, V, W)$:

$$\rho(U_t + U_x U + U_y V + U_z W) = (\sigma_{xx})_x + (\sigma_{xy})_y + (\sigma_{xz})_z + f_1$$
$$\rho(V_t + V_x U + V_y V + V_z W) = (\sigma_{yx})_x + (\sigma_{yy})_y + (\sigma_{yz})_z + f_2$$
$$\rho(W_t + W_x U + W_y V + W_z W) = (\sigma_{zx})_x + (\sigma_{zy})_y + (\sigma_{zz})_z + f_3 \qquad (6.3)$$

The vector (f_1, f_2, f_3) is the external force (per unit volume) acting on the fluid, while the stress terms represent, as explained in Section 5.1, the internal forces that the surrounding fluid exerts on our moving piece of fluid. But the stress–strain relations for a fluid are different than for a solid (5.4); they are now

$$\sigma_{xx} = -P + 2\mu U_x$$
$$\sigma_{yy} = -P + 2\mu V_y$$
$$\sigma_{zz} = -P + 2\mu W_z$$
$$\sigma_{xy} = \sigma_{yx} = \mu(U_y + V_x)$$
$$\sigma_{xz} = \sigma_{zx} = \mu(U_z + W_x)$$
$$\sigma_{yz} = \sigma_{zy} = \mu(V_z + W_y) \qquad (6.4)$$

If we insert the relations (6.4) into (6.3) and simplify using the divergence equation $U_x + V_y + W_z = 0$, we get the momentum equations of (6.1). For the time-dependent problem (6.1), the initial velocity (U, V, W) must be specified throughout the body at $t = 0$.

6.2 Stream Function and Penalty Method Approaches

The Navier–Stokes equations (6.1) are difficult to solve in their original form because of the first-order divergence equation and because only first derivatives of the pressure appear. For 2D problems,

$$\rho(U_t + U\,U_x + V\,U_y) = f_1 - P_x + \mu(U_{xx} + U_{yy})$$
$$\rho(V_t + U\,V_x + V\,V_y) = f_2 - P_y + \mu(V_{xx} + V_{yy})$$
$$U_x + V_y = 0 \tag{6.5}$$

a "stream function" approach can be used that makes these numerically more tractable. It can be shown that the fact that the divergence $U_x + V_y$ of the fluid velocity is zero guarantees that there is a "stream function" $\phi(x, y)$ such that $(U, V) = (\phi_y, -\phi_x)$, and then the divergence equation $U_x + V_y = \phi_{yx} - \phi_{xy} = 0$ is automatically satisfied.

Now let us define the "vorticity" by $\omega \equiv U_y - V_x = \phi_{xx} + \phi_{yy}$. Differentiate the first equation of (6.5) with respect to y, and the second with respect to x, and subtract, and we find that the pressure terms disappear and that we are left with the equations

$$\rho\omega_t + \rho(\phi_y\omega_x - \phi_x\omega_y) + (f_2)_x - (f_1)_y = \mu(\omega_{xx} + \omega_{yy})$$
$$\omega = \phi_{xx} + \phi_{yy} \tag{6.6}$$

where the second equation is just the definition of the vorticity. Now we have two second-order equations for the two unknowns ϕ and ω, and the pressure has been eliminated.

Another way to ensure that the divergence equation $U_x + V_y + W_z = 0$ is (nearly) satisfied is to replace the pressure in the first three equations of (6.1) by $P = -\alpha(U_x + V_y + W_z)$, where α is a very large number (compared with μ). This is called the penalty method, and it works for 3D problems also, and as discussed in Section 6.1, we are essentially just requiring that the fluid be *almost* incompressible.

6.3 Fluid Flow Examples

Example 6.1 (Stream function example) In this example we will find the steady-state flow (so the time derivative terms are zero) in a 2D pentagon with vertices at $(-1, -1), (1, -1), (1, 1), (0, 0.2), (-1, 1)$. We will assume an

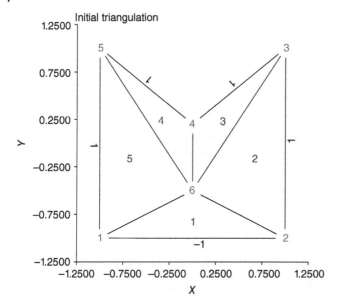

Figure 6.1 Initial triangulation, Example 6.1.

external force $\mathbf{f} = (y, -x)$, which tends to rotate the fluid around the origin. On the bottom of the pentagon, we will apply "free-slip" boundary conditions, $V = 0$, $U_y = 0$, and on the other four sides, we will apply "no-slip" boundary conditions, $U = 0$, $V = 0$.

We will first use a stream function approach (6.6), and we can verify (Problem 1a,b) that the free-slip conditions are $\phi = 0$, $\omega = 0$ and that the no-slip conditions are equivalent to setting ϕ and its normal derivative $\frac{\partial \phi}{\partial n}$ to 0.

We solved (6.6), with $\rho = 1.1$, $\mu = 0.1$, and Figure 6.2 shows a contour plot of the stream function. The contours of the stream function are parallel to the velocity field, because the gradient of ϕ is normal to a level curve of ϕ, and since $(\phi_x, \phi_y) \cdot (U, V) = \phi_x \phi_y - \phi_y \phi_x = 0$, it is also normal to the velocity. Because of the geometry, we have to use the Galerkin method (the region can be parameterized, but not smoothly). The initial triangulation used is shown in Figure 6.1; a uniform final triangulation of 1000 quadratic triangles was generated automatically by refining the initial triangulation.

Next we increase ρ to 22 and resolve the problem; this results in flow with a larger Reynold's number (i.e. the nonlinear terms in the Navier–Stokes equations are more dominant). Figure 6.3a shows the new velocity field, $(\phi_y, -\phi_x)$.

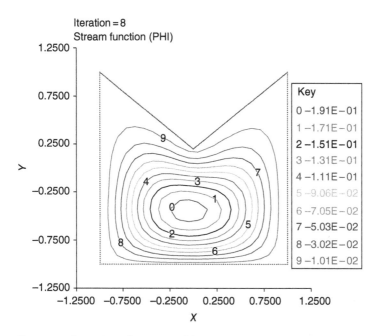

Figure 6.2 Streamlines (lower Reynold's number), Example 6.1.

Example 6.2 (Penalty method example) We also solved the problem with the higher Reynold's number using a penalty method formulation, that is, the 2D steady-state version of (6.3) and (6.4):

$$\rho(U_t + U\,U_x + V\,U_y) = [2\mu U_x - P]_x + [\mu(U_y + V_x)]_y + f_1$$
$$\rho(V_t + U\,V_x + V\,V_y) = [\mu(U_y + V_x)]_x + [2\mu V_y - P]_y + f_2$$
$$U_x + V_y = 0$$
$$\text{penalty method:} \quad P = -\alpha(U_x + V_y) \tag{6.7}$$

The free-slip conditions on the bottom ($V = U_y = 0$ and thus $U_y + V_x = 0$ also) are imposed by setting $GB1 = 0, GB2 = \text{zero}(V)$. Since $\text{zero}(V) \equiv \beta V$, where β is a large number chosen by PDE2D and $(N_x, N_y) = (0, -1)$ on the bottom, these mixed boundary conditions there are then

$$A1 * N_x + B1 * N_y = -\mu(U_y + V_x) = GB1 = 0$$
$$A2 * N_x + B2 * N_y = -(2\mu V_y - P) = GB2 = \beta V$$

The velocity field (Figure 6.3b) was very similar to that calculated using the stream function approach. With α in the range $10^3 - 10^{12}$, the integral of vorticity (0.448) changes very little and is close to that obtained using a stream function. If α is set too high, however, the problem can become ill conditioned.

(a)

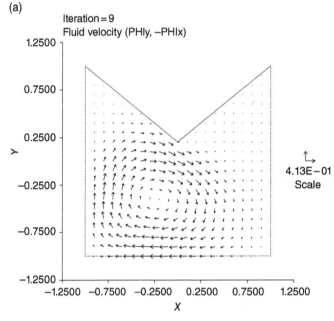

(b)

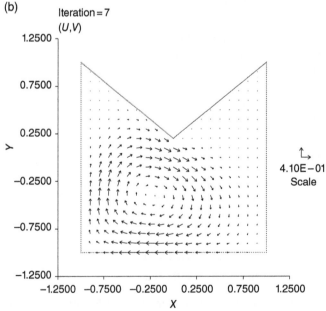

Figure 6.3 Higher Reynold's number velocity field. (a) Using stream function, Example 6.1 and (b) using penalty method, Example 6.2.

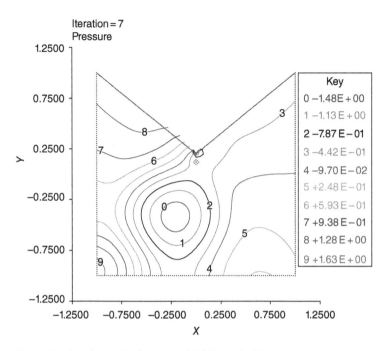

Figure 6.4 Penalty method pressure field, Example 6.2.

With the penalty method (but not the stream function approach), it is possible (sometimes!) to recover the pressure, using $P = -\alpha(U_x + V_y)$. Figure 6.4 shows a contour plot of the pressure field.

Example 6.3 (Fluid flow with heat convection) Next, we added the steady-state version of the heat conduction/convection equation (3.3), with $\kappa = 0.5, q = 0, C_p = 1$, with the temperature set to 100 on the bottom and 80 on the other sides. That is, we solved the fluid flow equations (6.7) (without the time derivatives, and with $\rho = 22, \mu = 0.1$ still) coupled with

$$(\kappa T_x - \rho C_p U\, T)_x + (\kappa T_y - \rho C_p V\, T)_y = 0$$

The resulting temperature distribution is shown in Figure 6.5.

Example 6.4 (Fluid flow in a torus) Finally, we solved the full 3D steady-state Navier–Stokes equations (6.1) in a hollowed-out torus of the same dimensions as the half torus of Figure I.2 (also used in Problem 4 of Chapter 3), but the whole torus this time (Figure 6.6). The penalty method was used to replace the

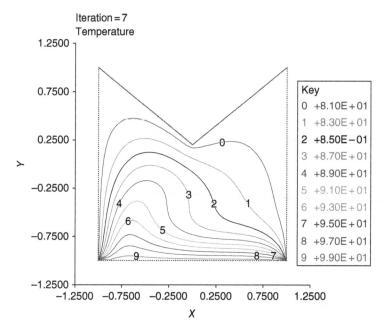

Figure 6.5 Coupled fluid flow and heat convection, Example 6.3.

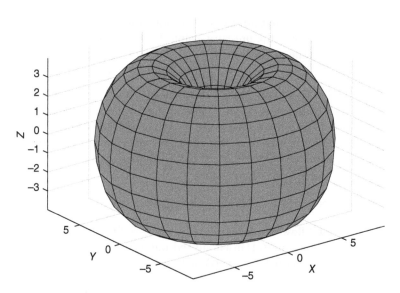

Figure 6.6 Torus of Example 6.4.

pressure by $P = -\alpha(U_x + V_y + W_z)$. The parametric equations (I.1), repeated here for convenience, were again used to define the torus:

$$X = (5 + p3 * cos(p2)) * cos(p1)$$
$$Y = (5 + p3 * cos(p2)) * sin(p1)$$
$$Z = p3 * sin(p2)$$

where $p1$ and $p2$ are the toroidal and poloidal angles and $p3$ is the distance from centerline. Now $0 \leq p1 \leq 2\pi, 0 \leq p2 \leq 2\pi, 1 \leq p3 \leq 4$, and there are periodic boundary conditions on both $p1$ and $p2$. At $p3 = 1$ (the inner core boundary), no-slip boundary conditions $U = V = W = 0$ were used, while at the outer boundary of the torus, $p3 = 4$, free-slip conditions were applied. At a general boundary point with unit outward normal vector $\mathbf{n} \equiv (N_x, N_y, N_z)$, free-slip boundary conditions mean that $UN_x + VN_y + WN_z = 0$ and that $(\frac{\partial U}{\partial n}, \frac{\partial V}{\partial n}, \frac{\partial W}{\partial n})$ must be parallel to $\mathbf{n}$. The first equation ensures that the velocity component $p \equiv UN_x + VN_y + WN_z$ normal to the boundary must be zero. The second requirement ensures that if $q \equiv UM_1 + VM_2 + WM_3$ is any velocity component tangential to the boundary, that is, parallel to any unit vector (M_1, M_2, M_3) that is perpendicular to $\mathbf{n}$, then $\frac{\partial q}{\partial n} = (\frac{\partial U}{\partial n}, \frac{\partial V}{\partial n}, \frac{\partial W}{\partial n}) \cdot (M_1, M_2, M_3) = 0$, since $(\frac{\partial U}{\partial n}, \frac{\partial V}{\partial n}, \frac{\partial W}{\partial n})$ is parallel to $\mathbf{n}$ and thus perpendicular to (M_1, M_2, M_3). Problem 1c shows that this means the traction components tangential to the boundary are zero.

Thus free-slip boundary conditions mean that

$$N_y \frac{\partial U}{\partial n} = N_x \frac{\partial V}{\partial n}$$
$$N_z \frac{\partial V}{\partial n} = N_y \frac{\partial W}{\partial n}$$
$$N_x \frac{\partial W}{\partial n} = N_z \frac{\partial U}{\partial n}$$
$$UN_x + VN_y + WN_z = 0 \tag{6.8}$$

The first three equations actually represent only two boundary conditions, because if $N_x \neq 0$, the two equations involving N_x imply the third (multiply the first equation by N_z/N_x and the third by N_y/N_x, and you get the second equation) and similarly if $N_y \neq 0$ or $N_z \neq 0$. We implemented the free-slip conditions by choosing a unit normal vector component with absolute value larger than $\frac{1}{2}$ (they can't all be smaller than this) and imposing the two equations involving that nonzero component. In the PDE2D boundary condition definitions, "UNORM" can be used to represent $\frac{\partial U}{\partial n}$.

We took $\mu = 1, \rho = 0.1, \alpha = 2000\mu$, and $(f_1, f_2, f_3) = (z, 0, 0)$, so there is an external force in the x direction that is positive above the $z = 0$ plane and negative below it. This causes a clockwise rotation of the fluid at the $p1 = 0$ cross section as seen in Figure 6.7a.

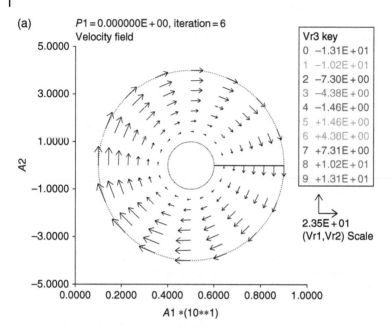

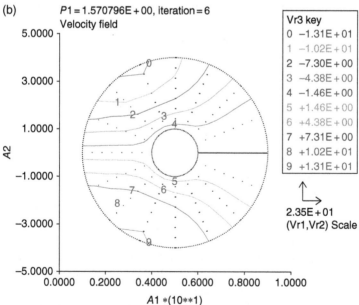

Figure 6.7 Velocity field cross sections, Example 6.4. (a) (U, W) with contour of V, at $p1 = 0$ and (b) (V, W) with contour of $-U$, at $p1 = \pi/2$.

To get the velocity field plots of Figure 6.7 at constant $p1$ cross sections, we had to set (in PMOD8Z)

$$UPRINT(1) = cos(p1) * U + sin(p1) * V$$
$$UPRINT(2) = -sin(p1) * U + cos(p1) * V$$

which means that when we ask for vector plots of (U, W) with a contour plot of V superimposed, U and V will actually be replaced by the above expressions. So at $p1 = 0$, there will be a vector plot of the in-plane velocity components (U, W) with a contour plot of the out-of-plane component V, while at $p1 = \pi/2$ a vector plot of (V, W) will be made with a contour plot of $-U$, as appropriate for this torus cross section. The axes for the cross-sectional plots were $A1 = \sqrt{x^2 + y^2}, A2 = z$.

Large values of α seem to cause serious ill-conditioning problems for the collocation method, and this makes solving the 3D Navier–Stokes equations with penalty method somewhat challenging for PDE2D. In fact, with large α the sparse direct method (ISOLVE=1) failed, and we had to use the parallel band solver option (ISOLVE=6), which solves the original equations rather than the more ill-conditioned normal equations (see Section B.4). The frontal method (ISOLVE=2) should work also, but it is slow.

6.4 Problems

1 (Free-slip boundary conditions)
   a) On a straight 2D boundary segment, with unit normal vector (N_x, N_y), let $p = (U, V) \cdot (N_x, N_y)$ be the velocity component normal to the boundary, and let $q = (U, V) \cdot (N_y, -N_x)$ be the component tangential to the boundary. Show that, using the stream function approach, if we set $\phi = C, \omega = 0$ along this boundary, this will enforce the "free-slip" boundary conditions $p = \frac{\partial q}{\partial n} = 0$. (Hints: Since ϕ is constant along the boundary, its tangential derivative $\frac{\partial \phi}{\partial s} = (\phi_x, \phi_y) \cdot (N_y, -N_x)$ must be zero, thus $p = 0$. Then show that $\frac{\partial q}{\partial n} - \frac{\partial p}{\partial s} = (q_x, q_y) \cdot (N_x, N_y) - (p_x, p_y) \cdot (N_y, -N_x) = \omega$.)
   b) Show that if we set $\phi = C, \frac{\partial \phi}{\partial n} = 0$ on a boundary, this will enforce "no-slip" boundary conditions $U = V = 0$.
   c) On a flat 3D boundary, show that the free-slip conditions given in (6.8) imply that the boundary "tractions" vector (g_1, g_2, g_3) defined by

$$\begin{bmatrix} g_1 \\ g_2 \\ g_3 \end{bmatrix} \equiv \begin{bmatrix} \sigma_{xx} & \sigma_{xy} & \sigma_{xz} \\ \sigma_{yx} & \sigma_{yy} & \sigma_{yz} \\ \sigma_{zx} & \sigma_{zy} & \sigma_{zz} \end{bmatrix} \begin{bmatrix} N_x \\ N_y \\ N_z \end{bmatrix}$$

with σ_{ij} given by (6.4) is parallel to the unit normal vector $\mathbf{n} \equiv (N_x, N_y, N_z)$, and so the traction components tangential to the boundary are zero.

Outline of demonstration: The vector $(\frac{\partial U}{\partial n}, \frac{\partial V}{\partial n}, \frac{\partial W}{\partial n})$ can be written as $J\mathbf{n}$ where J is the Jacobian matrix:

$$J \equiv \begin{bmatrix} U_x & U_y & U_z \\ V_x & V_y & V_z \\ W_x & W_y & W_z \end{bmatrix}$$

Thus the boundary conditions (6.8) ensure that $J\mathbf{n}$ is parallel to the unit normal vector $\mathbf{n}$. Now according to (6.8), the normal velocity component $p \equiv UN_x + VN_y + WN_z$ is zero along the boundary, so that boundary is a level surface for p, which means its gradient, which is $\nabla p = J^T\mathbf{n}$, is normal to its level surface and thus also parallel to $\mathbf{n}$. Now show that $(g_1, g_2, g_3) = [\mu(J + J^T) - PI]\mathbf{n}$ (P = pressure) so the tractions vector is parallel to $\mathbf{n}$.

Furthermore, show if the boundary conditions (6.8) are applied to an elasticity problem instead of a fluid flow problem, this will ensure that the boundary force vector (g_1, g_2, g_3) defined in (5.5), with stresses defined in (5.4), is parallel to the unit normal. Thus (6.8) is another (simpler) way to enforce the rolling friction (free-slip) boundary condition of Problem 1 of Chapter 5. Hint: You need to show that now $(g_1, g_2, g_3) = [c(J + J^T) + b(U_x + V_y + W_z)I]\mathbf{n}$, where $c = \frac{E}{2(1+v)}, b = \frac{Ev}{(1+v)(1-2v)}$.

d) On a straight 2D free-slip boundary arc, only the first and last of the 3D conditions (6.8) apply:

$$N_y \frac{\partial U}{\partial n} = N_x \frac{\partial V}{\partial n}$$
$$UN_x + VN_y = 0$$

and thus

$$N_y(U_xN_x + U_yN_y) = N_x(V_xN_x + V_yN_y)$$

For 2D problems, the fact that $J^T\mathbf{n}$ is parallel to $\mathbf{n}$, as shown in part (c), means that

$$N_y(U_xN_x + V_xN_y) = N_x(U_yN_x + V_yN_y)$$

Use these two results to confirm what was shown in part (a) that the vorticity $U_y - V_x$ must be zero on a straight free-slip boundary. Thus the boundary conditions on a 2D free-slip boundary can be taken as $UN_x + VN_y = 0, U_y - V_x = 0$.

e) If the velocity/displacement field for a fluid flow/elasticity problem on one side of a boundary is a mirror reflection of the field on the other side, we call this a "symmetry" boundary. Argue that a symmetry boundary condition is the same as a free-slip condition. (Hint: If p is the normal component of velocity/displacement, and q is a tangential component, what would happen if either p or $\frac{\partial q}{\partial n}$ were nonzero?)

2 (Time-dependent problem using stream function) Solve, using the Galerkin method, the time-dependent stream function equations (6.6) in the region of Examples 6.1–6.3, starting with initial conditions $U = V = \phi = \omega = 0$, but with free-slip conditions ($\phi = \omega = 0$) on the bottom and sides this time but still no-slip conditions (see Problems 1a and b) on the V-shaped top. Use $\rho = 10, \mu = 0.1, \mathbf{f} = (y, -x)$. Plot the velocity vector ($\phi_y, -\phi_x$) out to about $t = 40$, by which time a steady state (Figure 6.8) will have been reached. (Hint: To plot ($\phi_y, -\phi_x$) you can set APRINT(1)=PHIy, BPRINT(1)=-PHIx, assuming ϕ is named "PHI," then plot $(A1, B1)$.)

3 (Fluid flow with free-slip boundary)
   a) Use the collocation method to solve the steady-state 2D fluid flow equations (6.5), with penalty method formulation, and $\mu = 0.1, \rho = 10, \mathbf{f} = (-y, x)$, in the trapezoid of Figure 6.9a. This means set $P_x = -\alpha(U_{xx} + V_{yx})$ and $P_y = -\alpha(U_{xy} + V_{yy})$. This trapezoid can be parameterized by $x = p1, y = p2 * (1 + p1), 0 \le p1 \le 1, 0 \le p2 \le 1$, and boundary conditions are no slip on the left and right sides and free slip on the top (see Problem 1d) and bottom. Take $\alpha = 10^5 \mu$; higher values make the problem ill conditioned. Calculate the integral of $U^2 + V^2$, which should be about 0.0131. Pressure plots may be noisy even when the velocity field is accurate.
   b) Resolve the problem of part (a) using the Galerkin method and the steady-state equations (6.7), with penalty method formulation. Again calculate the integral of $U^2 + V^2$, to compare with part (a), and plot the pressure $P = -\alpha(U_x + V_y)$ (Figure 6.9b). The free-slip boundary conditions on the top can be written

$$GB1 = \text{zero}(U * NORMx + V * NORMy)$$
$$GB2 = \text{zero}(Uy - Vx)$$

The penalty parameter α can be much larger for the Galerkin method before the problem becomes too ill conditioned.
   c) Resolve the problem of part (a) using the collocation method and the steady-state stream function equations (6.6). Recall that in Problem 1a it was shown that free-slip conditions for the stream function approach can be modeled by $\phi = \omega = 0$ and in Problem 1b it was shown that no-slip conditions are $\frac{\partial \phi}{\partial n} = \phi = 0$.

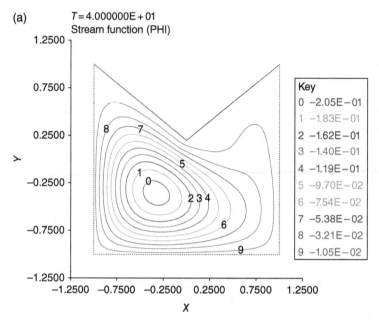

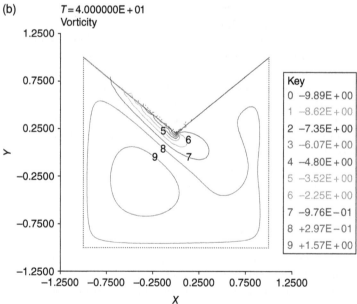

Figure 6.8 Problem 2, after steady-state reached. (a) Streamlines and (b) vorticity.

(a)
Iteration = 6
Velocity field

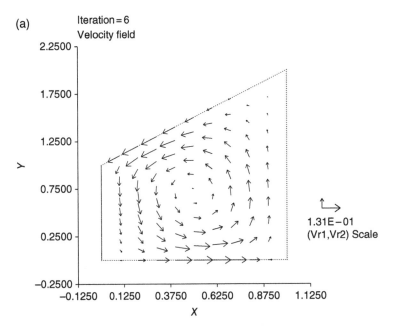

(b)
Iteration = 5
Pressure

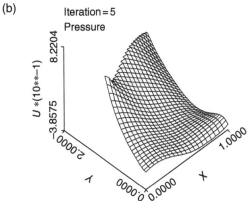

Figure 6.9 Penalty method formulation, Problems 3a and b. (a) Velocity field, using collocation method and (b) pressure, using Galerkin method.

Plot the streamlines (Figure 6.10a) and velocity field ($\phi_y, -\phi_x$), and calculate the integral of $U^2 + V^2 = \phi_y^2 + \phi_x^2$ to compare with parts (a) and (b).

4 (Fluid flow with inlet and outlet) Solve the 2D steady-state equations (6.7) with penalty method (Galerkin method) in the region of Examples 6.1–6.3,

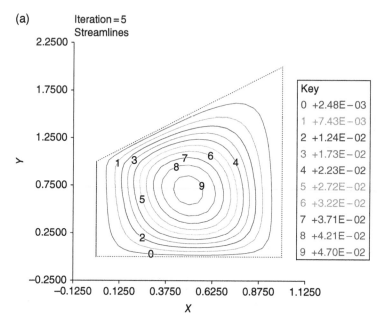

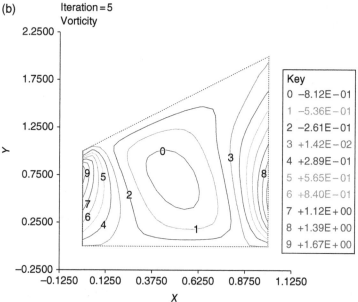

Figure 6.10 Stream function formulation, Problem 3c. (a) Streamlines and (b) vorticity.

but now with no external forces. There are again free-slip conditions ($V = 0, U_y = 0$, and thus $U_y + V_x = 0$) at the bottom and no-slip conditions ($U = V = 0$) on the V-shaped top; at the left boundary fluid is entering ($U = 1, U_y + V_x = 0$), and at the right (outlet) boundary, there are zero tractions, that is, GB1=GB2=0. Example 6.2 shows how to handle mixed boundary conditions like those on the bottom and left edges. Plot the velocity field (U, V) and the pressure, $P = -\alpha(U_x + V_y)$, as in Figure 6.11. Use $\rho = 10, \mu = 0.1$.

5 (Axisymmetric fluid flow problem) The axisymmetric Navier–Stokes equations are

$$\rho(U_t + U\,U_r + W\,U_z) = (\sigma_{rr})_r + (\sigma_{rz})_z + \frac{2\mu(U_r - U/r)}{r} + f_r$$

$$\rho(W_t + U\,W_r + W\,W_z) = (\sigma_{zr})_r + (\sigma_{zz})_z + \frac{\mu(U_z + W_r)}{r} + f_z$$

$$U_r + W_z + U/r = 0 \tag{6.9}$$

with

$$\sigma_{rr} = -P + 2\mu U_r$$
$$\sigma_{zz} = -P + 2\mu W_z$$
$$\sigma_{rz} = \sigma_{zr} = \mu(U_z + W_r) \tag{6.10}$$

where U and W are the velocity components in the r and z directions and f_r, f_z are the external forces in the r and z directions. These equations are derived in Problem 8.

Use the Galerkin method and the penalty method to replace the pressure by $P = -\alpha(U_r + W_z + U/r)$ (with $A1 = \sigma_{rr}, B1 = A2 = \sigma_{rz}, B2 = \sigma_{zz}$), and solve the steady-state axisymmetric equations in the hourglass figure shown in Figure 6.12. On the curved boundary, $r = 1 + z^2$, set $U = W = 0$, and $r = 0$ (the center of the hour glass) is a symmetry boundary (see Problem 1e), so we set $U = W_r = 0$ (thus also $GB2 = -\mu(U_z + W_r) = 0$) there. At the inlet ($z = -1$) there is a vertical inflow $W = 1 - (r/2)^2$ and $U_z + W_r = 0$, which can be imposed by setting GB1 = 0, GB2 = zero(W $- (1-(x/2)**2)$). At the outlet ($z = 1$), set GB1=GB2=0, which means the tractions are zero. Use $\mu = 0.02, \rho = 1, f_r = f_z = 0$ and plot the velocity field. INTRI=2 is highly recommended for constructing the initial triangulation, and of course r, z will have to be renamed x, y. Observe the back flow in Figure 6.13, when we zoom in on the upper right corner of the region.

Note that this is the Galerkin method, penalty formulation, version of interactive driver Example 13, which is solved there with the collocation method, and pressure as a third variable. Thus if you request the integral of $U + W$, you should get the same answer (3.52) as output by example 13.

(a)

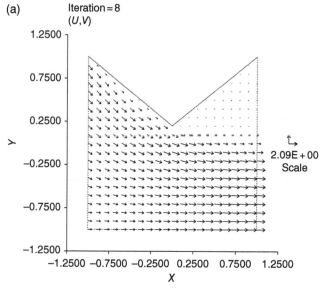

(b)

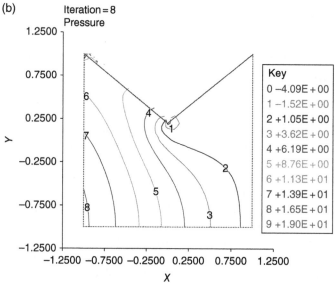

Figure 6.11 Problem 4. (a) Velocity field and (b) pressure.

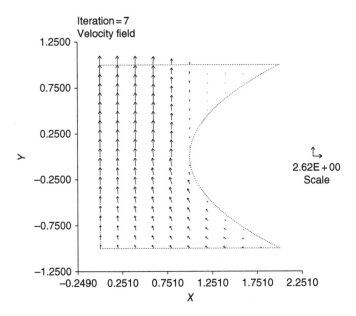

Figure 6.12 Axisymmetric flow, Problem 5.

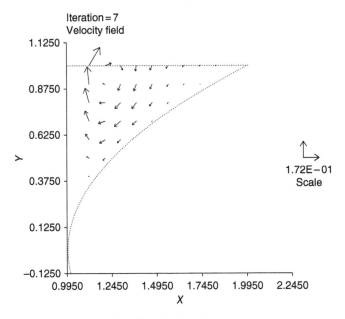

Figure 6.13 Zoom of back flow, Problem 5.

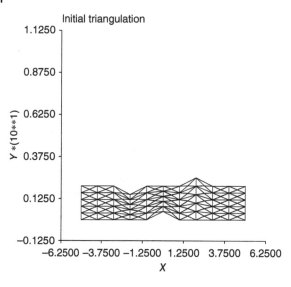

Figure 6.14 Automatically generated initial triangulation of Problem 6.

6 (Irrotational flow) Consider the steady-state stream function equations (6.6) in the pipe shown in Figure 6.14. Let's assume there are no external forces ($f_1 = f_2 = 0$), and let's set boundary conditions $U = 1$ and $\sigma_{xy} = U_y + V_x = 0$ on the inlet ($x = -5$) and outlet ($x = 5$) and free-slip conditions on the top and bottom. As shown in Problem 1a, free-slip conditions can be imposed by setting $\phi = C, \omega = 0$. Explain why the inlet and outlet conditions can be imposed by setting $\phi = y, \omega = 0$ there. But now we have $\omega = 0$ on the entire boundary, and with no nonhomogeneous terms, it is clear from (6.6) that the vorticity ω will be zero in the entire region. Thus we only need to solve Laplace's equation $\phi_{xx} + \phi_{yy} = 0$ for ϕ, so use PDE2D to solve this problem and plot the vector field ($\phi_y, -\phi_x$) (see Figure 6.15). When the vorticity is zero, this is called "irrotational" flow. Remember that ϕ must be constant on the top and bottom: For it to be continuous, what constants must be used?

Because of the geometry you must use the Galerkin method. You could set up the initial triangulation by hand (INTRI=3), but it is easier to set it up using the option INTRI=2, with $-5 \le p \le 5, 0 \le q \le 1$ and ($X(p,q), Y(p,q)$) defined as below:

```
if (p < -3) then
      y0 = 0
      y1 = 2
   else if (p < -2) then
      y0 = 0
      y1 = 2 - 0.5*(p+3)
   else if (p < -1) then
```

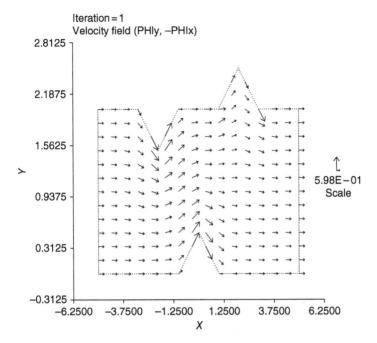

Figure 6.15 Irrotational flow, Problem 6.

```
    y0 = 0
    y1 = 1.5 + 0.5*(p+2)
else if (p < 0) then
    y0 = 0.5*(p+1)
    y1 = 2
else if (p < 1) then
    y0 = 0.5*(1-p)
    y1 = 2
else if (p < 2) then
    y0 = 0
    y1 = 2 + 0.5*(p-1)
else if (p < 3) then
    y0 = 0
    y1 = 2.5 - 0.5*(p-2)
else
    y0 = 0
    y1 = 2
endif
X = p
Y = y0 + q*(y1-y0)
```

This will *not* work for the collocation method, as the collocation method parameterizations must be continuously differentiable. They only need to be continuous for the Galerkin method, but now it is essential to have PGRID points at the corners $X = p = -3, -2, -1, 0, 1, 2, 3$.

7   (3D fluid flow) Solve the 3D steady-state Navier–Stokes equations (6.1) with penalty method $P = -\alpha(U_x + V_y + W_z)$ in the unit cube and with no-slip $(U = V = W = 0)$ boundary conditions at $x = 0, y = 0, z = 0$ and free-slip at $x = 1, y = 1, z = 1$. According to (6.8) free slip at $x = 1$, for example, where $\mathbf{n} = (1, 0, 0)$, means $U = V_x = W_x = 0$. Take $\rho = 1.1, \mu = 0.1$, and set an external force field $(f_1, f_2, f_3) = (-z, z, x - y)$. Since the region is rectangular, you can set ITRANS=0, and then p1,p2,p3 will represent X,Y,Z. Use $\alpha = 10^4$ or 10^5.

Plot the fluid velocity at cross sections $x = 0.5, 1$; choose ITPLOT=0, and make vector plots of (V, W) with the out-of-plane component U shown in a superimposed contour plot. Figure 6.16 shows a PDE2D velocity field plot at $x = 0.5$ and a MATLAB plot of pressure $(P = -\alpha(U_x + V_y + W_z))$ at the same cross section coded by color.

Figure 6.17 is a plot made by MATLAB function "streamline," which attempts to convey an idea of the entire velocity field in one picture. Whether or not this attempt is successful, it illustrates that it is easy to modify PDE2D-created MATLAB programs to generate alternative types of MATLAB graphics. The streamline plot was made by adding the following lines to the MATLAB program automatically generated by PDE2D:

```
figure
%  set starting points for streamlines on z=0.5 plane
[sx,sy,sz] = meshgrid(0.2:0.2:0.8,0.2:0.2:0.8,0.5);
streamline(X,Y,Z,U(:,:,:,NSAVE+1,1,1),U(:,:,:,NSAVE+1,1,2), ...
    U(:,:,:,NSAVE+1,1,3),sx,sy,sz,[0.05,380])
axis([xmin xmax ymin ymax zmin zmax])
xlabel('X'); ylabel('Y'); zlabel('Z')
```

8   (Derivation of axisymmetric Navier–Stokes equations) Derive the axisymmetric fluid flow equations (6.9) and (6.10) from the 3D equations (6.3) and (6.4) plus the divergence equation. Start with the fact that $U = R \cos(\theta)$, $V = R \sin(\theta)$, where R is the velocity component in the r direction, and remember that $R(r, z)$ and $W(r, z)$ are functions of r and z only. Multiply the first equation of (6.3) by $\cos(\theta)$ and the second by $\sin(\theta)$, and add the two, to give the first equation of (6.9); the third equation of (6.3) will reduce to the second equation of (6.9) after simplification.

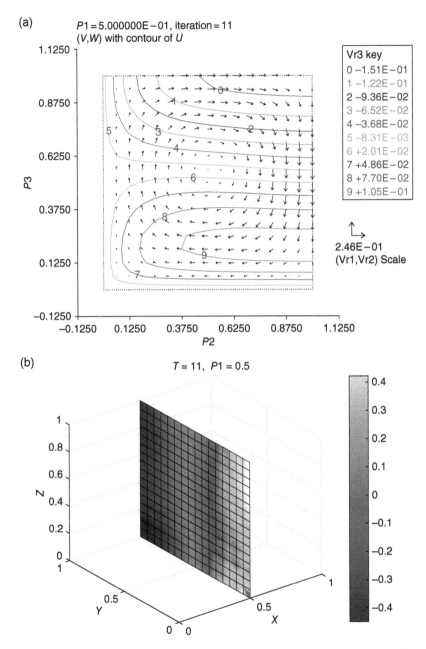

Figure 6.16 X=0.5 cross section, Problem 7. (a) Fluid velocity and (b) MATLAB plot of pressure.

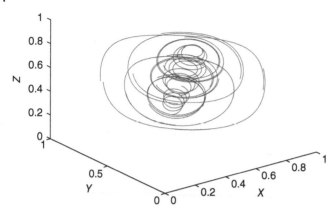

Figure 6.17 MATLAB streamline plot, Problem 7.

You should use the chain rule formulas given in Problem 4a of Chapter 5 repeated below:

$$U_x = R_r cos^2(\theta) + R\ sin^2(\theta)/r$$
$$U_y = R_r sin(\theta)cos(\theta) - R\ sin(\theta)cos(\theta)/r$$
$$U_z = R_z cos(\theta)$$
$$V_x = R_r sin(\theta)cos(\theta) - R\ sin(\theta)cos(\theta)/r$$
$$V_y = R_r sin^2(\theta) + R\ cos^2(\theta)/r$$
$$V_z = R_z sin(\theta)$$
$$W_x = W_r cos(\theta)$$
$$W_y = W_r sin(\theta)$$
$$W_z = W_z$$

Reducing the 3D divergence equation $U_x + V_y + W_z = 0$ to the axisymmetric divergence equation $R_r + W_z + R/r = 0$ of (6.9) is trivial using these formulas (of course, R is renamed to U in (6.9) and (6.10)), and deriving the left-hand sides of the first two equations of (6.9) is relatively easy also. The right-hand sides *would be* much more difficult to process, but notice that the right-hand sides of the 3D fluid flow equations (6.3) and (6.4) are the same as the right-hand sides of the 3D elasticity equations (5.2)/(5.4) with $E = 2\mu, v = 0$, except that the fluid flow equations have the extra terms involving the pressure, P. Thus you can use the result of Problems 4a and b of Chapter 5 and assume the right-hand sides of the axisymmetric fluid flow equations will be the same, except for the pressure terms, as the right-hand sides of the axisymmetric elasticity equations (5.10) and (5.11), with $E = 2\mu, v = 0$, and handle the pressure term separately, using $P_x cos(\theta) + P_y sin(\theta) = P_x x_r + P_y y_r = P_r$.

7

The Schrödinger and Other Eigenvalue Equations

7.1 The Schrödinger Equation

The N-particle Schrödinger equation has the form

$$-i\hbar \frac{\partial u}{\partial t} = \sum_{k=1}^{N} \frac{\hbar^2}{2m_k} \left[\frac{\partial^2 u}{\partial x_k^2} + \frac{\partial^2 u}{\partial y_k^2} + \frac{\partial^2 u}{\partial z_k^2} \right] - V(t, x_1, y_1, z_1, ...)u \qquad (7.1)$$

where $\hbar$ is Planck's constant, m_k is the mass of particle k, and V is the potential energy of the system. If u is the solution, $|u|^2$ represents the probability density for finding particle 1 at (x_1, y_1, z_1), particle 2 at (x_2, y_2, z_2), etc.

It does not seem to be possible to *derive* this equation from more basic principles. In *Partial Differential Equations* (Strauss 2008), Walter Strauss writes,

> Schrödinger's equation is most easily regarded simply as an axiom that leads to the correct physical conclusions, rather than as an equation that can be derived from simpler principles.... In principle, elaborations of it explain the structure of all atoms and molecules and so all of chemistry!

7.2 Schrödinger and Maxwell Equations Examples

The solution in (7.1) is a function of $3N + 1$ variables (including time), so PDE2D can solve only 1-particle problems ($N = 1$). If we look for solutions of the 1-particle time-dependent Schrödinger equation (7.1) that are periodic in time, that is, solutions of the form $u(x, y, z, t) = \phi(x, y, z)e^{-i\omega t}$, we get the steady-state Schrödinger equation:

$$\frac{\hbar^2}{2m} \nabla^2 \phi - V(x, y, z)\phi = -E\phi \qquad (7.2)$$

where $E = \hbar\omega$ is the energy of the particle.

Solving Partial Differential Equation Applications with PDE2D, First Edition. Granville Sewell.
© 2018 John Wiley & Sons, Inc. Published 2018 by John Wiley & Sons, Inc.

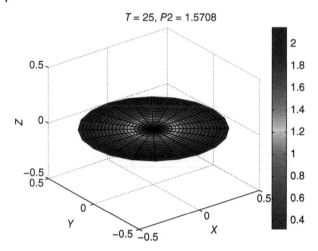

Figure 7.1 ϕ^2 for E_1 orbital in hydrogen atom.

Example 7.1 (The hydrogen atom) For the hydrogen atom, the simplest case, the potential energy of the electron is given by $V(\rho) = -\frac{e^2}{4\pi\epsilon_0\rho}$, where $\rho = \sqrt{x^2 + y^2 + z^2}$ is the distance of the electron from the more massive proton, which is considered fixed at the origin (hence this is really a 1-particle equation), e is the charge of an electron or proton, and ϵ_0 is the permittivity of free space. If the electron energy E is expressed in units of electron volts and distance in Angstroms, Eq. (7.2) becomes

$$\nabla^2\phi + \frac{3.7793}{\sqrt{x^2 + y^2 + z^2}}\phi = -0.26246\,E\phi \tag{7.3}$$

The boundary condition is that ϕ must be zero at $\rho = \infty$; we set $\phi = 0$ at $\rho = L_\infty = 16$ Angstroms. We are looking for energy values (E) for which there is a nonzero solution, so this is a 3D eigenvalue problem. We used spherical coordinates (ITRANS=2) with a nonuniform grid of 13 gridlines in the ρ direction (most dense near $\rho = 0$) and uniform grids of 9 lines each in the Φ and θ directions.

When we asked for the eigenvalue closest to -15, the value computed was -13.601. We made MATLAB plots of the probability distribution function ϕ^2, where ϕ is the corresponding eigenfunction, at constant longitude and constant latitude cross sections; one constant latitude cross section is shown in Figure 7.1.

When we change ITYPE to 4, PDE2D will find all eigenvalues (of the discrete problem, which approximate the smaller eigenvalues of (7.3)) without the eigenfunctions. For 3D problems this calculation is very time consuming, but it parallelizes well, and with the nonuniform grid described above (8424

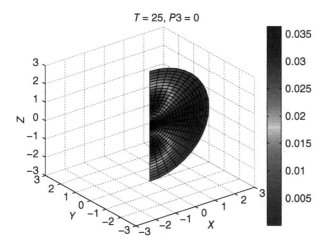

Figure 7.2 ϕ^2 for a nonspherically symmetric E_2 orbital.

unknowns), we were able to get all eigenvalues using 16 processors on a UTEP Centos cluster in 65 minutes. The first 14 were −13.601, −3.401, −3.401, −3.401, −3.398, −1.511, −1.511, −1.511, −1.511, −1.511, −1.511, −1.509, −1.509, and −1.507.

Notice there is one $E_1 = -13.60$ eigenvalue, four $E_2 = E_1/2^2 = -3.40$ values, and nine $E_3 = E_1/3^2 = -1.51$ values. As is known from solving (7.3) using separation of variables, one of each of these (the 1s, 2s, 3s orbitals) is spherically symmetric, that is, its eigenfunction depends on ρ only. In fact, by rerunning with ITYPE = 3 and setting EV0R very close to the eigenvalue desired, we were able to look at the eigenfunctions for −13.601, −3.398, and −1.507 and confirm that these (slightly different, in the discretized problem) eigenvalues were the spherically symmetric ones. Figure 7.2 shows a constant longitude cross section of the ϕ^2 corresponding to one of the nonspherically symmetric eigenfunctions of E_2.

Example 7.2 (Electron in a lattice) The Kronig–Penney problem (Patterson and Bailey 2007) models an electron in a crystal lattice and is based on a 1D version of the steady-state Schrödinger equation (7.2), with

$$V(x) = 0, \quad \text{for} \quad 0 < x < a - b$$
$$= \frac{\hbar^2 P}{mab}, \quad \text{for} \quad a - b < x < a$$

$V(x)$ is extended periodically, with period a, to all x.
If we define $\lambda = \frac{2mE}{\hbar^2}$, then (7.2) becomes

$$\phi_{xx} - W(x)\phi = -\lambda\phi \tag{7.4}$$

where

$$W(x) = 0, \quad \text{for} \quad 0 < x < a - b$$
$$= \frac{2P}{ab}, \quad \text{for} \quad a - b < x < a$$

It is known that the eigenfunctions $\phi(x)$ of a Schrödinger equation with periodic potential are not necessarily periodic with the same period but must have the form $\phi(x) = e^{ikx}U(x)$, where k is the wave number of this solution and $U(x)$ *is* periodic with the same period (a) as the potential. (And thus the probability $|\phi|^2 = |U|^2$ is also periodic.) Substituting this expression for ϕ into (7.4), we get

$$U_{xx} + 2ikU_x - k^2U - W(x)U = -\lambda U$$

We separate this complex equation into two real equations by defining $U = UR + i\,UI$:

$$UR_{xx} - 2kUI_x - k^2UR - W(x)UR = -\lambda\,UR$$
$$UI_{xx} + 2kUR_x - k^2UI - W(x)UI = -\lambda\,UI \tag{7.5}$$

Because of the periodic boundary conditions on UR, UI at $x = 0, a$, we had to use the collocation method. The discontinuity in $W(x)$ means (see (7.4)) that the second derivatives of the solution may be discontinuous, but that is not a problem for the collocation method (provided $a - b$ is a gridpoint), because its piecewise cubic polynomial approximations have continuous first, but not second, derivatives.

We found all eigenvalues (ITYPE=4) of (7.5), with $a = 1, b = 0.01, P = 1.5$ and $k = 0, 0.1\pi/a, 0.2\pi/a, ..., 2\pi/a$. Thus $W(x)$ has a large but narrow spike at the right end of the interval; in fact, for small b, $W(x) \approx \frac{2P}{a}\delta(x - a)$. Naturally, we used a nonuniform grid, with half of the 100 gridpoints placed in the small interval $(a - 2b, a)$. It is easy to solve with multiple values of k in one run, we just set NPROB=21, and the main DO loop was executed 21 times, with different values of k.

Patterson and Bailey show that the following relationship holds between the wave number k and the eigenvalues λ, when b is small:

$$cos(ka) = cos(\sqrt{\lambda}a) + Psin(\sqrt{\lambda}a)/(\sqrt{\lambda}a) \tag{7.6}$$

This was used to check our eigenvalues, and for all values of k, the first three eigenvalues all satisfied this relationship closely.

Figure 7.3 is a MATLAB plot of the first three eigenvalues $\lambda_1(k), \lambda_2(k), \lambda_3(k)$, which are proportional to the electron energies, against wave number k. Since $\lambda_1(k)$ can take any value from 2.389 to 9.870 (increasing or decreasing k by $2\pi/a$ will yield the same eigenvalue, as can be seen from (7.6)), and $\lambda_2(k)$ is always between 15.072 and 39.480, this means there are no eigenvalues for the original problem (7.4) between 9.870 and 15.072. There are also none between the highest second eigenvalue 39.480 and the lowest third, 45.214. And so the eigenvalues of the Kronig–Penney problem form continuous bands with gaps

Figure 7.3 First 3 eigenvalues, Kronig–Penney, Example 7.2.

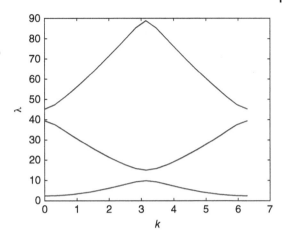

between the bands, and those gaps correspond to energy levels that the electron in the crystal cannot have.

Example 7.3 (Waveguide example) In Sewell and Cvetkovic (1989), Sewell (1989), Cvetkovic et al. (1994), and Zhao and Cvetkovic (1994), PDE2D was used to solve some waveguide equations, which are derived from Maxwell's equations and have the form

$$\frac{\partial}{\partial x}\left[\frac{\omega\mu H_x - \beta E_y}{\omega^2\mu\epsilon - \beta^2}\right] + \frac{\partial}{\partial y}\left[\frac{\omega\mu H_y + \beta E_x}{\omega^2\mu\epsilon - \beta^2}\right] + \omega\mu H = 0$$

$$\frac{\partial}{\partial x}\left[\frac{\omega\epsilon E_x + \beta H_y}{\omega^2\mu\epsilon - \beta^2}\right] + \frac{\partial}{\partial y}\left[\frac{\omega\epsilon E_y - \beta H_x}{\omega^2\mu\epsilon - \beta^2}\right] + \omega\epsilon E = 0 \tag{7.7}$$

with $\frac{\partial H}{\partial n} = E = 0$ on the boundary. Here μ is the permeability, ϵ is the permittivity, $\omega = 2\pi f$, where f is the wave frequency, and βi is the (purely imaginary) propagation constant. H and E are the magnetic and electric field components in the z direction (along the waveguide), from which the other components can be directly calculated.

If μ and ϵ are constants, Eqs. (7.7) reduce to an uncoupled ordinary eigenvalue PDE system for the eigenvalue β^2 :

$$\nabla^2 H + \omega^2\mu\epsilon H = \beta^2 H$$
$$\nabla^2 E + \omega^2\mu\epsilon E = \beta^2 E \tag{7.8}$$

But for waveguides, μ, ϵ are generally *not* constants.

For most values of β, the only solution to the homogeneous system (7.7) will be $H = E = 0$, but for certain values there will be nonzero solutions, which are not unique. Thus (7.7) is essentially an eigenvalue problem where the eigenfunction components H, E appear linearly; only the eigenvalue β appears nonlinearly.

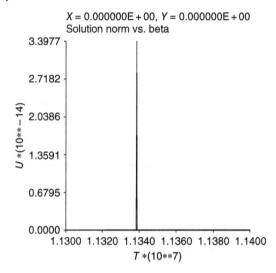

$X = 0.000000E + 00$, $Y = 0.000000E + 00$
Solution norm vs. beta

Figure 7.4 A sharp peak indicates an eigenvalue, Example 7.3.

We solve now a problem from Sewell and Cvetkovic (1989), which is the nonlinear eigenvalue problem (7.7) in a waveguide that consists of a 2-micron-by-2-micron glass square $(-a \leq x \leq a, -a \leq y \leq a, a = 10^{-6}m)$ embedded in an 8-micron-by-8-micron silica square $(-b \leq x \leq b, -b \leq y \leq b, b = 4 * 10^{-6}m)$. We use the Galerkin method, with gridlines along the material interfaces at $x = \pm a, y = \pm a$. The permeability of both materials is $\mu_0 = 12.566 * 10^{-7}$ Henries/m, and the permittivities are $\epsilon(x, y) = \epsilon_0 n^2 = 8.854 * 10^{-12} n^2$ Farads/m, where the refractive index for glass is $n = 1.55$ and for silica $n = 1.50$. The wave frequency is $f = 3.5269 * 10^{14}$/second.

Since PDE2D cannot solve nonlinear eigenvalue problems directly, we add some "random" nonzero terms $f_1(x, y) = sin(10x/a + 5y/a), f_2(x, y) = sin(5x/a + 10y/a)$ to the right-hand sides of the two equations in (7.7) and solve this nonhomogeneous system (with $\frac{\partial H}{\partial n} = E = 0$, and thus GB1=E=0, on the boundary) for a range of values of β. When β is close to an eigenvalue, the solution will be close to a large multiple of an eigenfunction.[1] So we calculate the integral of $|H| + |E|$ and plot this solution norm versus β and look for large peaks. The best way to vary β over a range of values is to solve (7.7), with the added nonhomogeneous terms, as a time-dependent problem,

---

1 This is best appreciated by looking at the discrete problem $(A - \beta I)\mathbf{x} = \mathbf{b}$. If β is very close to one eigenvalue of A, say, λ_k, and $A = SDS^{-1}$ is diagonalizable, the solution is $\mathbf{x} = S(D - \beta I)^{-1}S^{-1}\mathbf{b}$, and the k-th element on the diagonal matrix $(D - \beta I)^{-1}$ will be $\frac{1}{\lambda_k - \beta}$ which will be very large compared with the other diagonal elements. Thus $(D - \beta I)^{-1}S^{-1}\mathbf{b} \approx \frac{1}{\lambda_k - \beta} s_k \mathbf{e_k} \equiv \alpha \mathbf{e_k}$ where s_k is the k-th component of $S^{-1}\mathbf{b}$, assumed not to be zero, so α is large, and $(\mathbf{e_k})_j = \delta_{kj}$. Thus $\mathbf{x} \approx \alpha S \mathbf{e_k}$, where $S\mathbf{e_k}$ is the k-th column of S, which is the k-th eigenvector of A.

Figure 7.5 Eigenfunctions for Example 7.3. (a) *H* and (b) *E*.

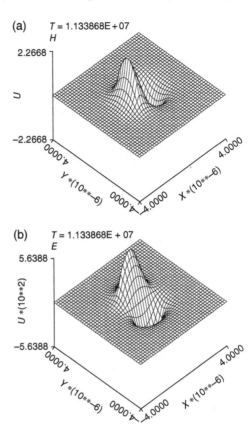

(a) $T = 1.133868E+07$
   H

(b) $T = 1.133868E + 07$
   E

and set $\beta = t$. Since there are no time derivatives, it is essential (see Example 1.2) to use the backward Euler method, so we set ADAPT=.FALSE. and CRANKN=.FALSE.. For this problem, an eigenvalue is known to exist between $\beta = 1.13 * 10^7$ and $1.14 * 10^7$, so we solve for $1.13 * 10^7 \leq t \leq 1.14 * 10^7$ and plot the solution norm as a function of time, and a sharp peak near the eigenvalue can be seen in Figure 7.4. (To create Figure 7.4, we set APRINT(1) = SINT(1) in PMOD8Z and requested a plot of A1 versus time.) Then we rerun, with the same grid, with a smaller time interval around the peak, and keep iterating until the large peak is located very accurately.

Once the value of $\beta_{peak} = 1.13386845 * 10^7$ in this range that gives the largest solution norm was located, rerunning, with exactly the same grid, with NSTEPS=NSAVE=1, $TF = \beta_{peak}$, resulted in the eigenfunction plots of H and E as shown in Figure 7.5. For reasonable eigenfunction plots, it is essential to have an accurate value of the eigenvalue; otherwise the eigenfunctions may be very noisy.

7.3 Problems

1 (1D hydrogen atom model)

a) Resolve Eq. (7.3) as a 1D problem, assuming ϕ is a function of ρ only, by writing $\nabla^2\phi$ as $\phi_{\rho\rho} + \frac{2}{\rho}\phi_\rho = \frac{1}{\rho^2}[\rho^2\phi_\rho]_\rho$.[2] You will only be able to find eigenvalues with eigenfunctions that depend on ρ only, but those include E_1, E_2, and E_3. For E_1 and E_2, plot $\rho^2\phi^2$ against $\rho(=x)$ in a line plot. (Hint: Set UPRINT(1) = X**2*PHI**2, then when you plot PHI, you are really plotting $\rho^2\phi^2$.) This will be proportional to the probability of the electron being a distance ρ from the proton, because the surface area of a sphere of radius ρ is $4\pi\rho^2$. For E_1, there should be a peak near $\rho = 0.5292$ Angstroms, the "Bohr radius" of hydrogen. Figure 7.6 shows the plot for E_2. Also calculate the expected value of the radius, $\int_0^\infty \rho(\rho^2\phi^2)\, d\rho / \int_0^\infty \rho^2\phi^2\, d\rho$, which will be larger than the radius of

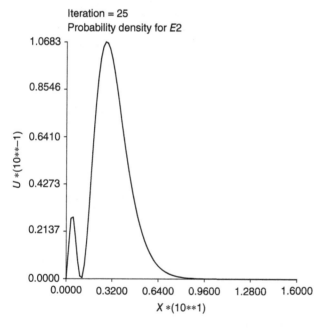

Figure 7.6 Probability density $\rho^2\phi^2$, for E_2, Problem 1a.

---

2 This is derived as follows: suppose ϕ is a function of $\rho \equiv \sqrt{x^2 + y^2 + z^2}$ only, then $\rho_x = \frac{x}{\rho}$ and $\phi_x = \phi_\rho\rho_x = \phi_\rho\frac{x}{\rho}$ and $\phi_{xx} = \phi_{\rho\rho}\rho_x\frac{x}{\rho} + \phi_\rho\frac{\rho - x\rho_x}{\rho^2} = \phi_{\rho\rho}\frac{x^2}{\rho^2} + \left[\frac{1}{\rho} - \frac{x^2}{\rho^3}\right]\phi_\rho$. Expanding ϕ_{yy} and ϕ_{zz} similarly and adding gives $\phi_{xx} + \phi_{yy} + \phi_{zz} = \phi_{\rho\rho}\frac{x^2+y^2+z^2}{\rho^2} + \left[\frac{3}{\rho} - \frac{x^2+y^2+z^2}{\rho^3}\right]\phi_\rho = \phi_{\rho\rho} + \frac{2}{\rho}\phi_\rho$.

maximum probability (the peak in your plot). For E_1 it will be exactly 1.5 times as large, 0.7938.

b) Now change ITYPE to 4 and compute all eigenvalues, without eigenfunctions (set p8z=-20 so the eigenvalues closest to -20 will be printed first). There will only be a few negative spherically symmetric eigenvalues (positive eigenvalues correspond to free electrons and are not of interest here) because the smaller negative eigenvalues have eigenfunctions that die down slowly and thus only show up when an ever larger L_∞ is used.

2   (The H_2 molecule)

a) Solve the Schrödinger equation for an electron under the influence of two protons (i.e. an H_2^+ molecule; see Fitzgerald and Sewell (2000)), located at $(0, 0, -R_0/2)$ and $(0, 0, R_0/2)$. That is, find the first (most negative) eigenvalue of

$$\frac{1}{2}(\phi_{xx} + \phi_{yy} + \phi_{zz}) + \left(\frac{1}{R_a} + \frac{1}{R_b}\right)\phi = -\lambda\phi, \qquad (7.9)$$

where

$$R_a = \sqrt{x^2 + y^2 + (z + R_0/2)^2}$$
$$R_b = \sqrt{x^2 + y^2 + (z - R_0/2)^2}$$
$$R_0 = 2.$$

The boundary condition is approximately imposed by setting $\phi = 0$ on the boundary of a large cube, $(-10, 10) \times (-10, 10) \times (-10, 10)$. Then use a nonuniform grid that is most dense near the two protons.

The eigenvalue λ represents the energy of the electron, and ϕ^2 is the electron's probability distribution function. The equation is in dimensionless form, and in this form the first three eigenvalues should be approximately $\lambda = -1.102$ (Flugge (1978) reports a value of -1.1017, after adjusting for the additional term of 0.5 in his V), -0.667, and -0.429. Remove all occurrences of C! in the Fortran program to enable creation of a MATLAB graphics program. Then run this using MATLAB to produce cross-sectional plots similar to Figure 7.1, which display the probability density ϕ^2 for the eigenfunction corresponding to the first eigenvalue.

Figure 7.7 shows the eigenfunctions (ϕ, not ϕ^2) corresponding to $\lambda = -1.102$ and $\lambda = -0.429$, using PDE2D's contour *surface* plotting algorithm (Sewell 1988).

b) The first two eigenvalues, but not the third, have axisymmetric eigenfunctions, so resolve (7.9) for these two, as an axisymmetric problem, that is, replace $\phi_{xx} + \phi_{yy} + \phi_{zz}$ by $\frac{1}{r}(r\phi_r)_r + \phi_{zz}$ (see Problem 4d of

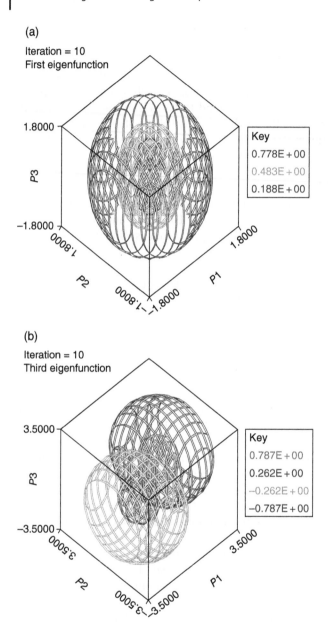

(a)
Iteration = 10
First eigenfunction

P3

1.8000

−1.8000

Key
0.778E + 00
0.483E + 00
0.188E + 00

P2

1.8000

−1.8000

P1

1.8000

−1.8000

(b)
Iteration = 10
Third eigenfunction

P3

3.5000

−3.5000

Key
0.787E + 00
0.262E + 00
−0.262E + 00
−0.787E + 00

P2

3.5000

−3.5000

P1

3.5000

−3.5000

Figure 7.7 H_2 molecule eigenfunctions, Problem 2a. (a) $\lambda = -1.102$ and (b) $\lambda = -0.429$.

Figure 7.8 Axisymmetric eigenfunctions, Problem 2b. (a) $\lambda = -1.102$ and (b) $\lambda = -0.667$.

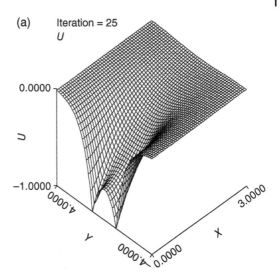

(a) Iteration = 25

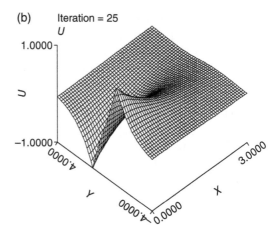

(b) Iteration = 25

Chapter 5) and solve in the rectangle $0 \leq r \leq 10, -10 \leq z \leq 10$. You can use either Galerkin or collocation, but Galerkin has the advantage that it is easier to grade the grid to make it dense near the two protons. If you use Galerkin, you can either request an adaptive triangulation or just set TRIDEN = 1/Ra + 1/Rb, where Ra and Rb are the distances to the two protons. Figure 7.8 shows both eigenfunctions (ϕ, not ϕ^2), zoomed in on the area around the protons.

Set $\phi = 0$ on $z = -10, z = 10, r = 10$, and no boundary condition (GB=0) at $r = 0$. And of course you must rename r to x and z to y.

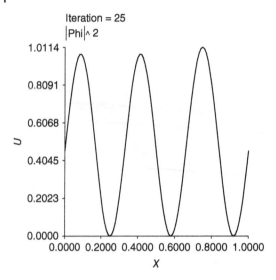

Iteration = 25
$|\text{Phi}|^{\wedge}2$

Figure 7.9 $|\phi|^2$ for third eigenfunction, with $b = 0.5$, Problem 3b.

3 (Kronig–Penney model)
   a) When $a = 1, b = 0.01, P = 1.5$, and $k = \pi$, it appears from Figure 7.3 that the third eigenvalue of the Kronig–Penney problem (7.5) is about $\lambda = 90$. Use a numerical method (Newton's method, for example) to solve the algebraic equation (7.6), where it is assumed that b is small, for a value of λ near 90.
   b) Use PDE2D to find the eigenvalue of (7.5) closest to $\lambda = 90$ (set EV0R=90) and the corresponding eigenfunction, with $a = 1, b = 0.01, P = 1.5$, and $k = \pi$, and make a plot of probability $|\phi|^2 = UR^2 + UI^2$. Your eigenvalue should be close to the value from part (a), since b is small. Repeat with $b = 0.5$ (Figure 7.9).

4 (Eigenvalues in cone) Find the smallest eigenvalue of

$$W_{xx} + W_{yy} + W_{zz} = \lambda W \quad \text{in the cone } 0 \le z \le 1 - r,$$
$$W = 0 \quad \text{on the boundary.}$$

You will need to set ITRANS=−3 and define an appropriate parameterization of the cone. (Answer: $\lambda = -36.92$.)
   Set NPROB=12 and EV0R = −30−5*IPROB, and make 12 runs to find the eigenvalues closest to $-35, -40, \dots - 90$. This "fishing expedition" should net 4 eigenvalues in this range. Look at the eigenfunction plots for each to see which are axisymmetric. One MATLAB cross-sectional plot for each of the first two eigenfunctions is shown in Figure 7.10, from which it appears that the first is axisymmetric while the second is not.

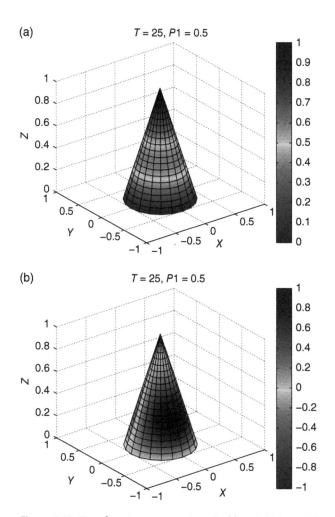

Figure 7.10 Eigenfunction cross sections, Problem 4. (a) $\lambda = -36.92$ and (b) $\lambda = -62.30$.

5 (Axisymmetric cone eigenvalues, Galerkin method) Two of the four eigenvalues found in Problem 4 have eigenfunctions that are functions of r and z only, so these two (but not all eigenvalues of the 3D problem) can also be found by solving the axisymmetric equation:

$$\frac{1}{r}\frac{\partial}{\partial r}\left(r\frac{\partial W}{\partial r}\right) + \frac{\partial}{\partial z}\left(\frac{\partial W}{\partial z}\right) = \lambda W \qquad 0 \leq r \leq 1, \qquad 0 \leq z \leq 1 - r.$$

Multiply this equation through by r, and find the smallest eigenvalue, using the Galerkin method. Note that since $A = rW_r$ and $B = rW_z$

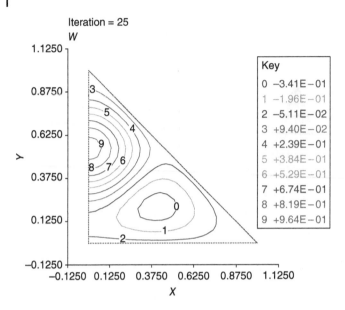

Figure 7.11 Second axisymmetric eigenfunction, Problem 6.

are 0 at $r = 0$, setting $GB = 0$ there is equivalent to "no" boundary condition.

Then change ITYPE to 4 with an editor, and rerun to find all eigenvalues of the axisymmetric problem. The two smallest should be the same as two of the four eigenvalues found in Problem 4. Finally, go back to ITYPE $= 3$, and find the eigenfunction for the second axisymmetric eigenfunction.

6 (Axisymmetric cone eigenvalues, collocation method) Repeat Problem 5 using the 2D collocation option. Figure 7.11 shows the second axisymmetric eigenfunction ($x \equiv r, y \equiv z$).

7 (Eigenvalues in parallelepiped) Find the smallest eigenvalue of $U_{xx} + U_{yy} + U_{zz}$ in a parallelepiped of Figure 7.12 with edges given by the vectors **A**=(2,1,1), **B**=(1,2,1), and **C**=(1,1,2), with $U = 0$ on the boundary. (Answer: $\lambda = -17.37$.) You can define this region using the parametric equations $(x, y, z) = p1 * \mathbf{A} + p2 * \mathbf{B} + p3 * \mathbf{C}$, where the parameters $p1, p2$, and $p3$ vary from 0 to 1. Make a contour plot of the eigenfunction in the plane $p3 = \frac{1}{2}$ with axes x and y. Also, remove all occurrences of $C!$ in the Fortran program to enable creation of a MATLAB graphics program, and run this using MATLAB.

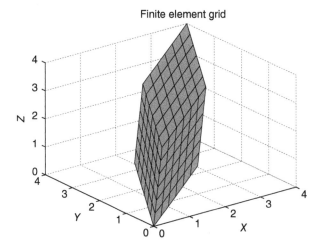

Figure 7.12 Parallelepiped of Problem 7.

8 (Complex eigenvalues)

   a) Use the GUI to create a program to solve the 1D eigensystem:

$$-U_{xx} - W_{xx} = \lambda\, U$$
$$U_{xx} - W_{xx} = \lambda W$$
$$U(0) = U(\pi) = 0$$
$$W(0) = W(\pi) = 0$$

   If you set the shift (EV0R) to 0 or any real number, you will find that the shifted inverse power method does not converge. So change ITYPE to 4 to get all the eigenvalues, and you will find they are all complex, $\lambda = n^2 \pm n^2 i$ for any integer n. (Eigenfunctions are $U = sin(nx)$, $W = \pm i\, sin(nx)$.) The shifted inverse power method cannot find complex eigenvalues of a real eigenvalue problem, with a real shift (EV0R), because the eigenvalues occur in complex conjugate pairs, equally distant from all real numbers (so there is no eigenvalue closest to EV0R to converge to).

   b) As seen in part (a), by setting ITYPE=4, you can still find the complex eigenvalues of a problem, but what if you want the eigenfunctions as well? Also ITYPE=4 is far more computationally expensive than the power method, especially for 2D and 3D problems. If you read the description of the parameter EVCMPX in your Fortran program, you will see that you *can* use the power method to find complex eigenvalues but only if you split your problem into real and imaginary parts:

$$-UR_{xx} - WR_{xx} = \lambda\, UR$$
$$-UI_{xx} - WI_{xx} = \lambda\, UI$$
$$UR_{xx} - WR_{xx} = \lambda WR$$
$$UI_{xx} - WI_{xx} = \lambda WI$$
$$UR(0) = UR(\pi) = 0$$
$$UI(0) = UI(\pi) = 0$$
$$WR(0) = WR(\pi) = 0$$
$$WI(0) = WI(\pi) = 0$$

where $U = UR + i\,UI$, $W = WR + i\,WI$.

Create a new PDE2D program (you cannot just change EVCMPX to .TRUE.) to solve this system of four equations. Set the complex shift

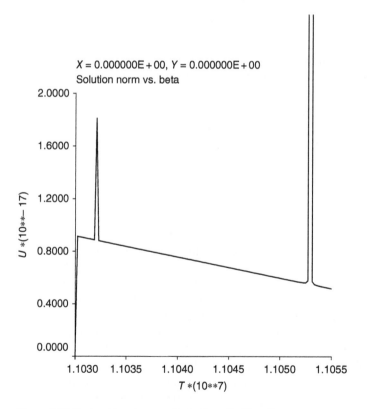

Figure 7.13 Peaks at two known eigenvalues, Problem 9.

(EV0R + i EV0I) to $8 + 7\ i$, and you should get convergence to the closest eigenvalue, which is $\lambda = 9 + 9\ i$. And now you can see the eigenfunction.

9 (Eigenvalues of nonlinear problem) Recreate the program for Example 7.3, but set n to 1.5 in both materials, then μ and ϵ are constant and so the system (7.7) reduces to the ordinary eigenvalue problem (7.8), which has eigenvalues at $\beta^2 = \omega^2 \mu\epsilon - (k^2 + l^2)(\frac{\pi}{2b})^2$ for all integers k, l. (What are the eigenfunctions?) Run this program with appropriate time intervals to verify that it can find the peaks corresponding to one or two of these eigenvalues. Solve the problem in the form (7.7), not (7.8), since the point is to verify that adding nonhomogeneous terms and looking for values of β that make the solution large can be used to find known eigenvalues of a nonlinear eigenvalue problem.

Figure 7.13 shows a run that locates the peaks at $\sqrt{\omega^2 \mu\epsilon - 5(\frac{\pi}{2b})^2} = 1.1052646 * 10^7$ and $\sqrt{\omega^2 \mu\epsilon - 8(\frac{\pi}{2b})^2} = 1.1031697 * 10^7$.

8

Minimal Surface and Membrane Wave Equations

8.1 Derivation of Minimal Surface Equation

To derive the minimal surface equation, suppose $u(x, y)$ is the height of a surface with $u = g(x, y)$ on the boundary $\partial\Omega$ of Ω, which minimizes the surface area (derived in most Calculus III texts):

$$SA(u) \equiv \iint_\Omega \sqrt{1 + u_x^2 + u_y^2} \, dA.$$

Then if $e(x, y)$ is any smooth function with $e = 0$ on the boundary, $SA(u + \alpha e) \geq SA(u)$ for any α; thus

$$f(\alpha) \equiv SA(u + \alpha e) = \iint_\Omega \sqrt{1 + (u_x + \alpha e_x)^2 + (u_y + \alpha e_y)^2} \, dA$$

should have a minimum at $\alpha = 0$ and so $\frac{df}{d\alpha}(0)$ should be zero. But

$$\frac{df}{d\alpha}(\alpha) = \iint_\Omega \frac{(u_x + \alpha e_x)e_x + (u_y + \alpha e_y)e_y}{\sqrt{1 + (u_x + \alpha e_x)^2 + (u_y + \alpha e_y)^2}} \, dA$$

so

$$\frac{df}{d\alpha}(0) = \iint_\Omega \nabla e \cdot \left[\frac{\nabla u}{\sqrt{1 + u_x^2 + u_y^2}} \right] dA$$

$$= \iint_\Omega \nabla \cdot \left[e\frac{\nabla u}{\sqrt{1 + u_x^2 + u_y^2}} \right] - e\nabla \cdot \left[\frac{\nabla u}{\sqrt{1 + u_x^2 + u_y^2}} \right] dA$$

$$= \int_{\partial\Omega} e\frac{\nabla u \cdot \mathbf{n}}{\sqrt{1 + u_x^2 + u_y^2}} \, ds - \iint_\Omega e\nabla \cdot \left[\frac{\nabla u}{\sqrt{1 + u_x^2 + u_y^2}} \right] dA$$

Solving Partial Differential Equation Applications with PDE2D, First Edition. Granville Sewell.
© 2018 John Wiley & Sons, Inc. Published 2018 by John Wiley & Sons, Inc.

where **n** is the unit outward normal. Here we have done a (multivariate) integration by parts (see Appendix A, formula (A.3)). Then, since $e = 0$ on $\partial \Omega$

$$\frac{df}{d\alpha}(0) = -\iint_{\Omega} e \nabla \cdot \left[\frac{\nabla u}{\sqrt{1 + u_x^2 + u_y^2}} \right] dA = 0$$

Since e is an arbitrary smooth function in the interior of Ω, we conclude that the factor multiplying e must be zero everywhere, which gives the minimal surface equation

$$\nabla \cdot \left[\frac{\nabla u}{\sqrt{1 + u_x^2 + u_y^2}} \right] = 0 \tag{8.1}$$

8.2 Derivation of Membrane Wave Equation

The rigid plates modeled in Chapter 2 resist bending; a membrane attached to a frame resists stretching, so it tries to minimize its surface area, and thus its height $u(x, y)$ satisfies (when it is in equilibrium) the minimal surface equation (8.1).

Now let's assume a tightly stretched membrane is subjected to a vertical force (per unit area) $f(x, y, t)$ and is displaced from equilibrium initially. There may also be a vertical frictional force $-bu_t$ proportional to the velocity and in an opposing direction. To derive an equation for the dynamic behavior of the membrane height $u(x, y, t)$, we apply Newton's second law to the vertical components of acceleration and force on a small section S of the membrane:

$$\iint_S \rho u_{tt} \, dA = \iint_S T \nabla^2 u \, dA + \iint_S [f - bu_t] \, dA \tag{8.2}$$

Here T is the membrane tension, and ρ is the density (mass per unit area) of the membrane, so the integral on the left is the mass of section S of the membrane times its vertical acceleration. The second integral on the right represents the external and frictional forces in the vertical direction, and the first integral term is the internal vertical force on S exerted by the rest of the membrane, as will be justified now.

The net vertical force that the rest of the membrane exerts on S at a point on the boundary ∂S will be in the direction of the vector $\mathbf{V} = (N_x, N_y, \frac{\partial u}{\partial n})$, where $\mathbf{n} = (N_x, N_y)$ is the unit outward normal to the boundary ∂S and will by definition have magnitude equal to the tension $T(x, y)$. If we assume the membrane gradients are small compared with 1 (we did not assume this in deriving the minimal surface equation), then $\| \mathbf{V} \| = \sqrt{N_x^2 + N_y^2 + (\frac{\partial u}{\partial n})^2} = \sqrt{1 + (\frac{\partial u}{\partial n})^2} \approx 1$, so $\mathbf{V}$ is a unit vector, and the force (per unit length) on the boundary point is given by $T\mathbf{V}$. Thus the net force that the rest of the membrane exerts on section S is the vector:

$$\int_{\partial S} T\mathbf{V}\, ds = \int_{\partial S} \left(TN_x, TN_y, T\frac{\partial u}{\partial n} \right) ds = \int_{\partial S} (T\mathbf{i} \cdot \mathbf{n}, T\mathbf{j} \cdot \mathbf{n}, T\nabla u \cdot \mathbf{n})\, ds =$$

$$\iint_S (\nabla \cdot (T\mathbf{i}), \nabla \cdot (T\mathbf{j}), \nabla \cdot (T\nabla u))\, dA = \iint_S (T_x, T_y, \nabla \cdot (T\nabla u))\, dA$$

where $\mathbf{i} = (1,0), \mathbf{j} = (0,1)$ and we have used the divergence theorem (formula (A.1)) on each component of the vector. Now, we are assuming the external and frictional forces are in the vertical direction only and that the membrane only moves vertically, so this internal force must also have no horizontal components. Therefore $T_x = T_y = 0$ and so the tension T is actually a constant[1] and the vertical component of this internal force is $\iint_S \nabla \cdot (T\nabla u)\, dA = \iint_S T\nabla^2 u\, dA$, which explains the appearance of this term in (8.2).

Since S is an arbitrary small subregion, (8.2) means that everywhere

$$\rho u_{tt} + b u_t = T\nabla^2 u + f \tag{8.3}$$

This is the (damped) membrane wave equation. In the absence of external forces, the steady-state (equilibrium) equation is $\nabla^2 u = 0$, which is what the minimal surface equation (8.1) reduces to when we assume, as we did in deriving the membrane wave equation, that the gradients are not large.

The potential energy of the membrane is the tension T times the surface area, so the total (kinetic plus potential) energy of the membrane in a region Ω is

$$E(t) = \iint_\Omega \frac{1}{2}\rho u_t^2 + T\sqrt{1 + \nabla u \cdot \nabla u}\, dA$$

But if we assume as before that the gradients of u are small compared with 1 and use the approximation $\sqrt{1 + \epsilon} \approx 1 + \frac{1}{2}\epsilon$, for small ϵ, then we can say that

$$E(t) \approx \iint_\Omega \left[\frac{1}{2}\rho u_t^2 + \frac{T}{2}\nabla u \cdot \nabla u \right] dA + C \tag{8.4}$$

Although $C = T\, area(\Omega)$, there is an arbitrary constant in the definition of potential energy; only changes in potential energy are significant, so we can reset C to 0 for convenience.

---

1 We assume the membrane is already tightly stretched at equilibrium, so small increases in surface area do not alter the tension, and thus T is not a function of time either.

If there is no external force adding energy to the system ($f = 0$) and if the membrane height is fixed on the boundary (so $u_t = 0$ there; alternatively, if $\frac{\partial u}{\partial n} = 0$),

$$E'(t) = \iint_{\Omega} \rho u_t u_{tt} + T\nabla u \cdot \nabla u_t \, dA$$

$$= \iint_{\Omega} \rho u_t u_{tt} + \nabla \cdot [u_t T\nabla u] - u_t \nabla \cdot [T\nabla u] \, dA$$

$$= \iint_{\Omega} u_t [\rho u_{tt} - T\nabla^2 u] \, dA + \int_{\partial\Omega} Tu_t \frac{\partial u}{\partial n} \, ds = \iint_{\Omega} -bu_t^2 \, dA$$

Thus without external forces, as shown for the spring problem in Section 1.1, the total energy is constant if there is no friction and decreases if there is (until a steady state has been reached, $u_t = 0$).

8.3 Examples

Example 8.1 (Minimal surface example) We solved the 2D nonlinear minimal surface problem (8.1) in the "hourglass" figure, $-1 \leq x \leq 1, -(1 + x^2) \leq y \leq (1 + x^2)$. On the boundary of this figure, the surface height is specified as $u = x^2/2$. We solved using the Galerkin method with the INTRI=2 initial triangulation generation option to simplify defining the initial triangulation and made PDE2D surface and contour plots of $u(x, y)$ (Figure 8.1). Since this is a highly nonlinear problem, we used the boundary condition function to provide a reasonable initial guess in the interior, $u(x, y) = x^2/2$.

Example 8.2 (Damped membrane wave example) Next we solved the membrane wave equation (8.3) in a parallelogram with edges given by the vectors (3,1) and (1,3) and one vertex at the origin. Since this equation is second order in time, we have to write it as a system of two first-order equations by defining $v \equiv u_t$:

$$u_t = v$$
$$\rho v_t = -bv + T(u_{xx} + u_{yy}) + f$$

The collocation method was used, and the region was defined using parametric equations $x = 3 * p1 + p2, y = p1 + 3 * p2, 0 \leq p1 \leq 1, 0 \leq p2 \leq 1$. The membrane is attached on its boundary to a frame with boundary conditions

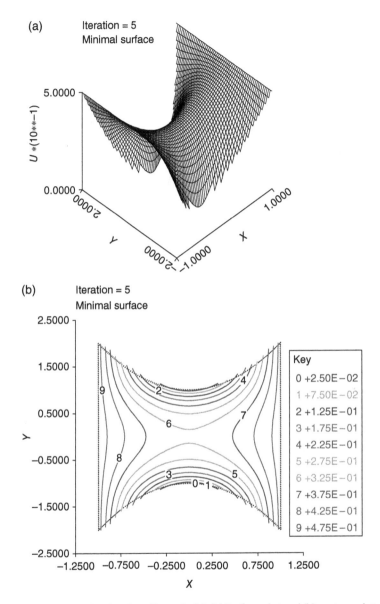

Figure 8.1 Minimal surface, Example 8.1. (a) Surface plot and (b) contour plot.

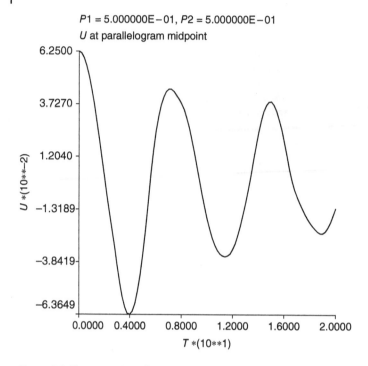

Figure 8.2 Time variation of U at parallelogram midpoint, Example 8.2.

$u = 0, v \equiv u_t = 0$ and initial conditions $u = p1(1 - p1)p2(1 - p2), v = 0,$[2] so the initial conditions match the boundary conditions at the boundary. Values of $\rho = 2, b = 0.2, T = 0.5, f = 0$ were used, and we integrated out to $t = 20$.

The height at the parallelogram midpoint is shown as a function of time in Figure 8.2; notice the damping of the oscillations, because $b > 0$.

Figure 8.3 shows the membrane at $t = 4$ and 8. The contour plot at $t = 4$ was made by replacing the call to the MATLAB function "surf," in the MATLAB program automatically generated by PDE2D, with calls to "contour" and "clabel."

8.4 Problems

1 (Minimal surface, collocation method) Resolve Example 8.1 using the collocation method. You will need to expand out the derivatives in (8.1),

---

2 Although for the collocation method the PDEs, boundary, and initial conditions can be, and usually are, written as functions of $x, y,$ and the derivatives of the unknowns with respect to $x, y,$ they can, if desired, be written as functions of $p1, p2,$ and the derivatives with respect to these parameters, and for the initial conditions here, this was more convenient.

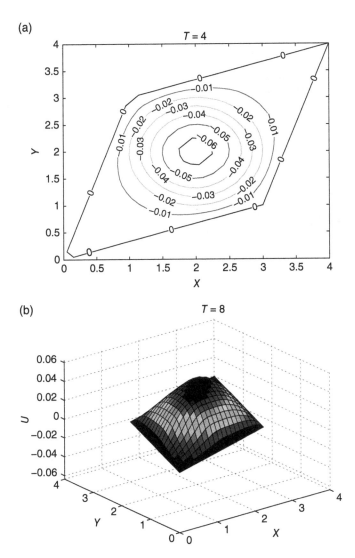

(a)

T = 4

(b)

T = 8

Figure 8.3 MATLAB plots of membrane, Example 8.2. (a) Contour plot at $t = 4$ and (b) surface plot at $t = 8$.

and you will need to parameterize the region, with constant limits on the parameters $p1, p2$ (hint: See I.2). To get convergence of Newton's method for this nonlinear problem, you may need to multiply the nonlinear terms by $\beta = min(1.d0, (t - 1)/5.)$, which means that at the first iteration ($t = 1$), we are solving a linear problem so initial values are not important and after five iterations we are solving the full nonlinear problem.

(a)

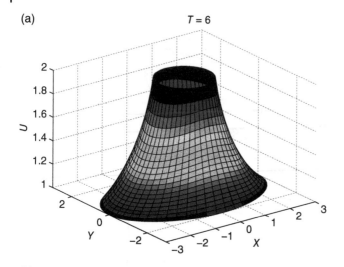

(b)

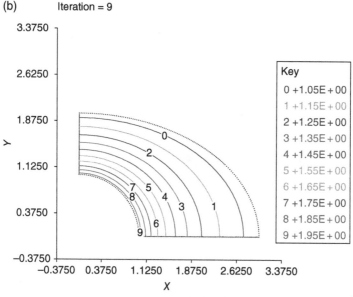

Figure 8.4 Minimal surface in annulus, Problem 2. (a) Full region (Problem 2a) and (b) symmetry used to reduce region (Problem 2b).

2 (Minimal surface in annulus)

    a) Solve the minimal surface PDE (8.1) in a region consisting of the ellipse $\frac{x^2}{3^2} + \frac{y^2}{2^2} \leq 1$, with a circular hole $x^2 + y^2 \leq 1$ removed (Figure 8.4a). Set $U = 2$ on the inner circle and $U = 1$ on the outer ellipse. Use the collocation method, compute the integral of U, and make a surface plot of the solution.

b) Resolve Problem 2a using the Galerkin method. You can solve the PDE in the first quadrant only since, by symmetry, $\frac{\partial U}{\partial n} = 0$ at $x = 0$ and $y = 0$. In this way you can get the same accuracy with one fourth as many elements. Make a contour plot of the solution. If you use the INTRI option 2, you can use the same parameterization as used in part (a), as $(X(p,q), Y(p,q))$, to generate the initial triangulation. The integral of $4U$ should be the same as the integral of U found in part (a).

3 (Membrane wave in parallelogram) Resolve Example 8.2 using the Galerkin method. You can define the initial triangulation of the parallelogram using either INTRI=2 or INTRI=3. If you use a constant step size and ISOLVE=1, 2, or 4, you can save an enormous amount of computer time by setting NOUPDT=.TRUE..

Let PDE2D calculate and output at each time step the total energy integral (8.4) (with $C = 0$) to see that it decreases as the wave is damped. Reset b to 0, and see if the energy is constant now (it will be if CRANKN=.TRUE.). Hint: To plot the total energy integral versus time (Figure 8.5b), set APRINT(1) = SINT(1) in PMOD8Z, and request a plot of A1 versus time at any point (X, Y).

4 (String wave problem) Solve the 1D problem (8.3), without the u_{yy} term or external forces, which models waves in a string:

$$\rho u_{tt} + b u_t = T u_{xx}$$

The string length is $L = 10$, and there are reflecting boundary conditions $u_x(0, t) = u_x(L, t) = 0$ and initial conditions $u(x, 0) = max(0, 1 - |x - 5|), u_t(x, 0) = 0$. Use $\rho = 2, b = 0.2, T = 0.5$. Again, reduce to a system of two first-order equations:

$$u_t = w$$
$$\rho w_t = -bw + T u_{xx}$$

Since $u_x = 0$ for all t at the endpoints, $u_{xt} = u_{tx} = 0$ so $w_x = 0$ also.

Figure 8.6 shows a plot of $u(x, t)$ as a function of both x and t. You can see how the initial "tent" separates into two parts, which spread to the boundaries, reflect back, and recombine at $t = 20$. The wave amplitude is damped by the frictional term $-bu_t$. Notice that the wave seems to travel at a speed of $10/20 = 0.5$, can you explain the speed? (Hint: Verify that $u(x, t) \equiv g(x \pm ct)$ is a solution of $u_{tt} = c^2 u_{xx}$ for any smooth function $g(x)$ and this solution is constant along lines $x = k \pm ct$.)

(a)

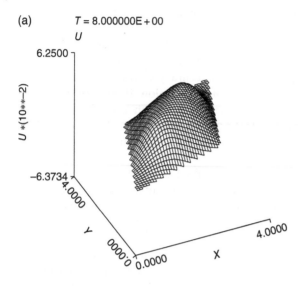

(b)

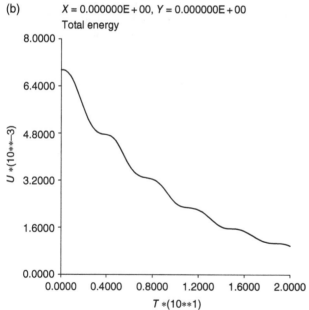

Figure 8.5 Damped membrane, Problem 3. (a) PDE2D plot of membrane at $t = 8$ and (b) total energy as function of t.

Figure 8.6 String height as function of x and t, Problem 4.

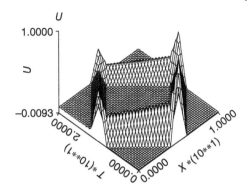

5 (Minimal curves) Derive the "minimal curve" equation, that is, the 1D version of the minimal surface equation, starting from the equation for arc length:

$$AL(u) \equiv \int_0^L \sqrt{1 + u_x^2} \, dx.$$

If $u(x)$ is the function with $u(0) = A, u(L) = B$, which minimizes this arc length integral, and $e(x)$ is any smooth function with $e(0) = e(L) = 0$ and $f(\alpha) \equiv AL(u + \alpha e)$, explain why $\frac{df}{d\alpha}(0) = 0$, and use this to find a differential equation satisfied by u. Then simplify this equation and say "oh yeah, of course!"

9

The KPI Wave Equation[1]

9.1 A Difficult Nonlinear Problem

In this final chapter, we use PDE2D to solve a very challenging nonlinear problem.[1]

The Kadomtsev–Petviashvili I (KPI) wave equation

$$U_{xt} + 6U_x^2 + 6UU_{xx} + U_{xxxx} = 3U_{yy} \tag{9.1}$$

is used to model waves in thin films with high surface tension. It has been extensively studied in the mathematical community since the paper by Kadomtsev and Petviashvili (Kadomtsev and Petviashvili 1970). We will solve this equation using PDE2D's Galerkin method, with initial conditions consisting of two-lump solitons, which collide and reseparate. Since the solution has steep, moving peaks, an adaptive finite element grid is used with a grading that moves with the peaks.

To use PDE2D, we have to reduce this fourth-order equation to a system of three first- or second-order equations by introducing the variables $V \equiv U_x, W \equiv U_{xx}$:

$$0 = U_x - V$$
$$0 = U_{xx} - W$$
$$V_t = -W_{xx} + 3U_{yy} - 6V^2 - 6UW$$

Lu, Tian, and Grimshaw (2004) give a two-lump soliton analytical solution (astonishing that an analytical solution can be found) of the KPI equation,

---

1 This chapter is based on an article (Sewell 2013) published in *Bulletin of Computational Applied Mathematics* (www.compama.co.usb.ve).

Solving Partial Differential Equation Applications with PDE2D, First Edition. Granville Sewell.
© 2018 John Wiley & Sons, Inc. Published 2018 by John Wiley & Sons, Inc.

expressed as $Q(x, y, t) = 2[\Phi\Phi_{xx} - \Phi_x^2]/\Phi^2$, where $\Phi(x, y, t)$ is defined as the determinant of a certain 4 by 4 matrix. We will use this analytical solution for defining initial conditions and to compare with our numerical solution.

Initial conditions for the problems are

$$U(x, y, 0) = Q(x, y, 0)$$
$$V(x, y, 0) = Q_x(x, y, 0)$$
$$W(x, y, 0) = Q_{xx}(x, y, 0)$$

Two of the problems solved in Lu et al. (2004) will be solved here:

1) An "oblique collision" problem, where two solitons of equal size collide at a 90° angle and pass through each other.
2) A "direct collision" problem, where two solitons are initially located along the x-axis, moving to the right with different velocities. The larger soliton overtakes the smaller one, and they combine and reseparate.

In both cases, as long as the two solitons are sufficiently separated initially, the initial conditions can be represented approximately by

$$Q(x, y, 0) \approx 16\frac{N_1}{D_1^2} + 16\frac{N_2}{D_2^2}$$

where, for the oblique collision case,

$$N_j = -4(x - x_j - 2k_j(y - y_j))^2 + 16k_j^2(y - y_j)^2 + 1/k_j^2$$

$$D_j = 4(x - x_j - 2k_j(y - y_j))^2 + 16k_j^2(y - y_j)^2 + 1/k_j^2$$

with $(x_1, y_1) = (15, -15), (x_2, y_2) = (15, 15), k_1 = \frac{1}{2}, k_2 = \frac{-1}{2}$

and for the direct collision case,

$$N_j = -4(x - x_j)^2 + 16k_j^2(y - y_j)^2 + 1/k_j^2$$

$$D_j = 4(x - x_j)^2 + 16k_j^2(y - y_j)^2 + 1/k_j^2$$

with $(x_1, y_1) = (15, 0), (x_2, y_2) = (31, 0), k_1 = \frac{\sqrt{6}}{4}, k_2 = \frac{\sqrt{6}}{8}$

In each case, $16\frac{N_j}{D_j^2}$ has a peak of $16k_j^2$ at (x_j, y_j), and since this term dies out at a distance, $U(x, y, 0)$ will have peaks close to (x_1, y_1) and (x_2, y_2), as seen in Figures 9.1a and 9.5a. In the oblique case, the solitons have velocities $(6, 6)$ and $(6, -6)$ and in the direct case, $(4.5, 0)$ and $(1.125, 0)$.

Figure 9.1 Oblique
collision, $t = 0$.
(a) $U(x, y, 0)$ and
(b) triangulation.

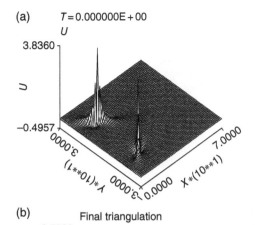

(a) $T = 0.000000E + 00$

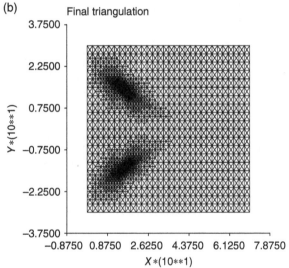

(b)   Final triangulation

The boundary conditions are chosen to reflect the fact that $Q(x, y, t)$ goes to zero far from the soliton peaks. Recall that the original problem (9.1) is fourth order in x and second order in y:

$$U(0, y, t) = 0$$
$$V(0, y, t) = 0$$
$$W(0, y, t) = 0$$
$$U_x(70, y, t) = 0$$
$$W_x(70, y, t) = 0$$
$$U_y(x, -30, t) = 0$$
$$U_y(x, 30, t) = 0$$

(a)

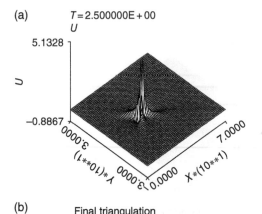

$T = 2.500000E + 00$

U

Figure 9.2 Oblique collision, $t = 2.5$. (a) $U(x, y, 2.5)$ and (b) triangulation.

(b)

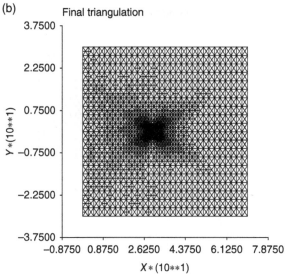

As seen in Figures 9.1–9.9, the solutions have steep, moving peaks, so a major challenge is constructing an appropriately graded, moving grid. PDE2D does not actually allow the triangular grid to change with time. However, an adaptive moving grid is improvised as follows: PDE2D is called multiple times, each time it solves the system from $t = t_{n-1}$ to $t = t_n$, taking several time steps, using the Crank–Nicolson method to discretize time, on a fixed grid, and the solution at t_n is dumped on a uniform (1000 by 1000) mesh. The next time PDE2D is called, it generates a new triangulation (of 4800 cubic triangular elements, with

Figure 9.3 Oblique collision, $t = 5.0$. (a) $U(x, y, 5.0)$ and (b) triangulation.

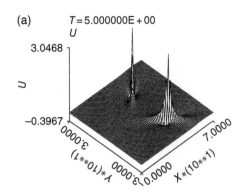

(a) $T = 5.000000E + 00$

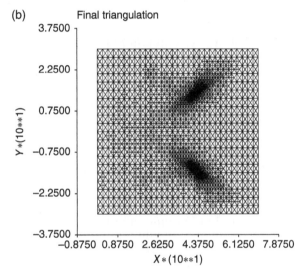

(b) Final triangulation

Figure 9.4 Oblique collision exact solution, $t = 5.0$.

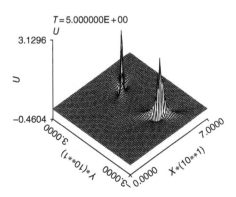

$T = 5.000000E + 00$

(a) $T = 0.000000E + 00$
U

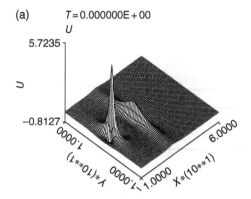

(b)   Final triangulation

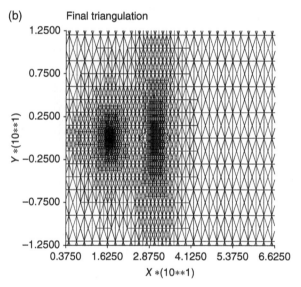

Figure 9.5 Direct collision, $t = 0$. (a) $U(x, y, 0)$ and (b) triangulation.

about 65 000 unknowns) adaptively, based on the final solution at t_n, with initial conditions linearly interpolated from the dumped solution at t_n. This process is done quite automatically, the call to PDE2D is simply placed inside a DO loop, with initial time t_{n-1} and final time t_n each call, and the dump/restart and adaptive triangulation options are turned on.

This improvised moving adaptive grid illustrates the claim made in the Introduction that PDE2D has "all the flexibility of Fortran" and shows why this is often very important in solving difficult problems.

Figure 9.6 Direct collision, $t = 5.0$. (a) $U(x, y, 5.0)$ and (b) triangulation.

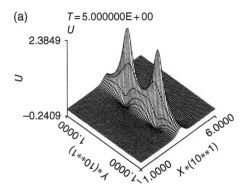

(a) $T = 5.000000E + 00$

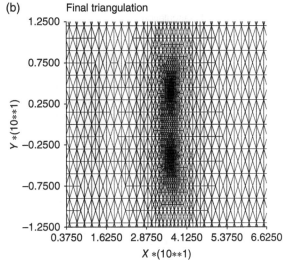

(b)   Final triangulation

9.2 Numerical Results

Results are shown in Figures 9.1–9.3 for the oblique collision case, where a time step of $\Delta t = 0.0125$ is used and the grid is updated adaptively every 10 steps. The moving grid follows the peaks very nicely, and the final solution, at t=5, agrees reasonably well with the analytic solution (compare Figures 9.3a and 9.4) at that time. In all problems, it is known that both the integral of U and the integral of U^2 should be constant with time. This gives us an easy way to estimate the numerical error; a norm of the actual error $|U - Q|$ was not used because with these sharp peaks, that error might be very large even when the approximate solution is good. At t=5, the error in the U integral was 97%, and the error

(a)

$T = 8.000000E + 00$

U

2.0779

U

−0.2038

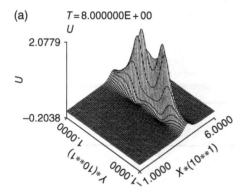

1.0000

Y*(10**1)

1.0000

1.0000

X*(10**1)

6.0000

Figure 9.7 Direct collision, $t = 8.0$. (a) $U(x, y, 8.0)$ and (b) triangulation.

(b)   Final triangulation

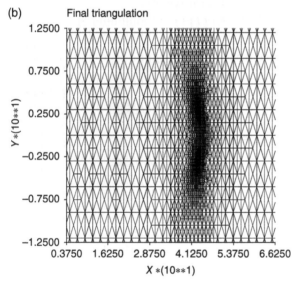

$X *(10**1)$

in the U^2 integral was 9.4%. The integral of U is much more sensitive than the integral of U^2 to the smaller values near the boundary, far from the peaks, so the problem was resolved with U, V, W set to the true solution on the entire boundary (note: these boundary conditions were *not* used for Figures 9.1–9.10), and the error in the U integral decreased to 3.3%, while the error in the U^2 integral increased slightly to 10.0%. Since U is not always positive, it may be more reasonable to divide by the integral of $|U|$ rather than the integral of U, in calculating the relative error; when this is done, we get a more respectable-looking figure of 0.7% for the error in the integral of U.

Figure 9.8 Direct collision, $t = 10.0$. (a) $U(x, y, 10.0)$ and (b) triangulation.

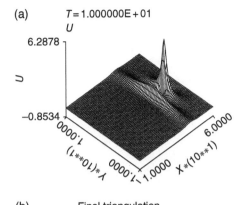

(a) $T = 1.000000E + 01$

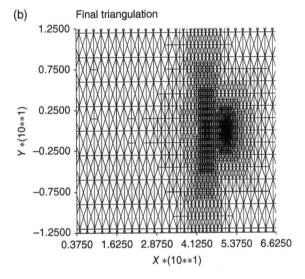

(b) Final triangulation

For the direct collision case, results are shown in Figures 9.5–9.8. Again the moving grid follows the peaks nicely, and the PDE2D solution agrees well with the analytic solution until about $t=5$, when the taller, faster peak catches the smaller one (compare Figures 9.6a and 9.9a). After that, the peaks computed by PDE2D separate more slowly than they should: The PDE2D solution at $t=10$ looks much like the true solution at $t=8$ (compare Figures 9.8a and 9.9b)!

For the direct collision problem, a time step of $\Delta t = 0.025$ was used, and again the grid was updated adaptively every 10 steps. The integral of U^2 differs from the true value at $t=10$ by about 5.9%.

(a)   T = 5.000000E + 00

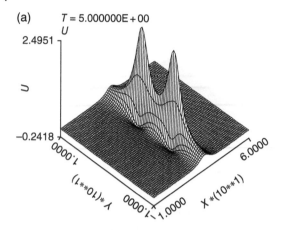

Figure 9.9 Direct collision exact solution, $t = 5.0$ and $t = 8.0$. (a) $Q(x, y, 5.0)$ and (b) $Q(x, y, 8.0)$.

(b)   T = 8.000000E + 00

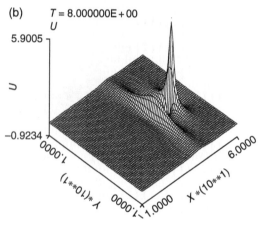

Finally, we resolved both problems with the same number of elements and same time step sizes but this time using a constant, uniform triangulation. The resulting solutions, shown in Figure 9.10a and b, are very bad and clearly illustrate the importance of the moving adaptive grid. The error in the integral of U^2 for the oblique collision problem at $t=2.5$ is now 500%, and 4000% at $t=5$ for the direct collision problem! Notice that the direct collision solution is not only

Figure 9.10 Results with uniform grid. (a) Oblique collision, $t = 2.5$ and (b) direct collision, $t = 5.0$.

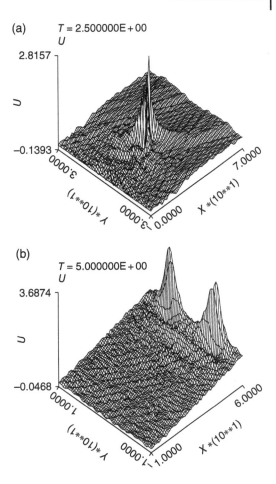

quite noisy, but also the peaks are very far from where they should be (compare Figures 9.9a and 9.10b). The fact that a uniform triangulation of 4800 cubic elements produces such a poor solution illustrates how difficult this nonlinear problem is.

Appendix A

Formulas from Multivariate Calculus

Here are some multivariate calculus formulas that were used in earlier chapters.

In what follows, u, w represent scalar functions, and $\mathbf{v} = (v_1, v_2, v_3)$ represents a vector function. $\mathbf{n}$ represents the unit outward normal vector to the boundary $\partial\Omega$ of Ω. The formulas are all still valid, with the obvious modifications, if Ω is a two-dimensional region:

$$\text{Gradient } u \equiv \nabla u = \left(\frac{\partial}{\partial x}, \frac{\partial}{\partial y}, \frac{\partial}{\partial z} \right) u = \left(\frac{\partial u}{\partial x}, \frac{\partial u}{\partial y}, \frac{\partial u}{\partial z} \right)$$

$$\text{Divergence } \mathbf{v} \equiv \nabla \cdot \mathbf{v} = \left(\frac{\partial}{\partial x}, \frac{\partial}{\partial y}, \frac{\partial}{\partial z} \right) \cdot (v_1, v_2, v_3) = \frac{\partial v_1}{\partial x} + \frac{\partial v_2}{\partial y} + \frac{\partial v_3}{\partial z}$$

$$\nabla \cdot \nabla u = \frac{\partial^2 u}{\partial x^2} + \frac{\partial^2 u}{\partial y^2} + \frac{\partial^2 u}{\partial z^2} \equiv \nabla^2 u$$

A.1. Divergence theorem:

$$\iiint_\Omega \nabla \cdot \mathbf{v} = \iint_{\partial\Omega} \mathbf{v} \cdot \mathbf{n}$$

A.1b. Divergence theorem applied to $\mathbf{v} = \nabla u$:

$$\iiint_\Omega \nabla^2 u = \iiint_\Omega \nabla \cdot \nabla u = \iint_{\partial\Omega} \nabla u \cdot \mathbf{n} = \iint_{\partial\Omega} \frac{\partial u}{\partial n}$$

A.2. Product rules (the second is just the first with $\mathbf{v} = \nabla w$):

$$\nabla \cdot (u\mathbf{v}) = u\nabla \cdot \mathbf{v} + \nabla u \cdot \mathbf{v}$$

$$\nabla \cdot (u\nabla w) = u\nabla^2 w + \nabla u \cdot \nabla w$$

Solving Partial Differential Equation Applications with PDE2D, First Edition. Granville Sewell.
© 2018 John Wiley & Sons, Inc. Published 2018 by John Wiley & Sons, Inc.

A.3. Integration by parts (follow from product rules A.2 and divergence theorem A.1)

$$\iiint_\Omega u\nabla \cdot \mathbf{v} = \iint_{\partial\Omega} u\mathbf{v} \cdot \mathbf{n} - \iiint_\Omega \nabla u \cdot \mathbf{v}$$

$$\iiint_\Omega u\nabla^2 w = \iint_{\partial\Omega} u\nabla w \cdot \mathbf{n} - \iiint_\Omega \nabla u \cdot \nabla w$$

$$= \iint_{\partial\Omega} u\frac{\partial w}{\partial n} - \iiint_\Omega \nabla u \cdot \nabla w$$

Appendix B

Algorithms Used by PDE2D

This appendix includes frequent references to "The Numerical Solution of Ordinary and Partial Differential Equations, Third Edition" (Sewell 2015), and "Computational Methods of Linear Algebra, Third Edition" (Sewell 2014), where many of the algorithms used are described in more detail.

B.1 The Galerkin and Collocation Finite Element Methods

By far the most widely used form of the finite element method is the Galerkin method, and it is one of the two finite element variants used by PDE2D. We will illustrate its use for the following simple 3D steady-state PDE and will discuss the time-dependent and eigenvalue problems later, in Sections B.6 and B.7:

$$-\nabla^2 u = f(x, y, z) \quad \text{in } R,$$
$$u = r(x, y, z) \quad \text{on } \partial R \tag{B.1}$$

We try to find an approximate solution of the form

$$U(x, y, z) = \Omega(x, y, z) + \sum_{i=1}^{M} a_i \Phi_i(x, y, z), \tag{B.2}$$

where $\{\Phi_1, \ldots, \Phi_M\}$ is a set of linearly independent basis functions that vanish on the boundary ∂R and Ω is another function that satisfies the boundary condition $\Omega = r$ on ∂R. Clearly, U will satisfy the boundary condition regardless of the values chosen for $a_1, \ldots, a_M$.

Then we multiply both sides of (B.1) by Φ_k and integrate over R

$$\iiint_R -\nabla^2 U \, \Phi_k \, dx \, dy \, dz = \iiint_R f \Phi_k \, dx \, dy \, dz$$

Solving Partial Differential Equation Applications with PDE2D, First Edition. Granville Sewell.
© 2018 John Wiley & Sons, Inc. Published 2018 by John Wiley & Sons, Inc.

Using a multivariate integration by parts (A.3) and the fact that $\Phi_k = 0$ on the boundary ∂R gives

$$\iiint_R \nabla U \cdot \nabla \Phi_k \, dx \, dy \, dz - \iint_{\partial R} \frac{\partial U}{\partial n} \Phi_k dA = \iiint_R f \Phi_k \, dx \, dy \, dz$$

$$\iiint_R \nabla U \cdot \nabla \Phi_k \, dx \, dy \, dz = \iiint_R f \Phi_k \, dx \, dy \, dz$$

When we replace U by its expansion (B.2),

$$\iiint_R \left[\nabla \Omega + \sum_{i=1}^M a_i \nabla \Phi_i \right] \cdot \nabla \Phi_k \, dx \, dy \, dz = \iiint_R f \Phi_k \, dx \, dy \, dz$$

we get a system of M linear equations for the M unknown parameters $a_1, \ldots, a_M$:

$$\sum_{i=1}^M A_{ki} a_i = b_k \quad \text{for } k = 1, \ldots, M \tag{B.3}$$

where

$$b_k = \iiint_R [f \Phi_k - \nabla \Omega \cdot \nabla \Phi_k] dx \, dy \, dz$$

$$A_{ki} = \iiint_R \nabla \Phi_k \cdot \nabla \Phi_i \, dx \, dy \, dz \tag{B.4}$$

The finite element choices for the basis functions $\Phi_1, \ldots, \Phi_M$ are piecewise polynomial functions that are each zero outside some small subregion of R, and Ω is usually a linear combination of similar basis functions. This ensures that A is sparse, since $A_{ki} = 0$ unless the regions where Φ_k and Φ_i are nonzero overlap. When there are more general boundary conditions than in the simple problem (B.1), there will be a boundary integral also in the formulas for A_{ki}, b_k; see section 5.1 of Sewell (2015). The integrals required by A_{ki}, b_k have to be calculated using some numerical quadrature formulas, of course.

The collocation method is the other brand of the finite element method that is used by PDE2D. Here the approximate solution U may again have the form (B.2), and the Φ_i may be the same type of piecewise polynomial functions as for the Galerkin method, except that for the collocation method, they must have continuous first derivatives when solving a second-order PDE, while the Galerkin method does not require this. Now the unknown coefficients $a_1, \ldots, a_M$ are found by requiring that the approximate solution satisfy the PDE (B.1) exactly at M collocation points (x_k, y_k, z_k):

$$-\nabla^2 U(x_k, y_k, z_k) = f(x_k, y_k, z_k)$$

$$-\sum_{i=1}^{M} a_i \nabla^2 \Phi_i(x_k, y_k, z_k) = f(x_k, y_k, z_k) + \nabla^2 \Omega(x_k, y_k, z_k)$$

A linear system of the form (B.3) again results, with

$$b_k = f(x_k, y_k, z_k) + \nabla^2 \Omega(x_k, y_k, z_k)$$
$$A_{ki} = -\nabla^2 \Phi_i(x_k, y_k, z_k) \tag{B.5}$$

The matrix is again sparse, because Φ_i and its derivatives are nonzero only in a small subregion, and thus are zero at most of the collocation points. When there are more general boundary conditions, those boundary conditions are enforced exactly at some boundary collocation points; see section 5.4 of Sewell (2015) for an example. Note that the fact that the collocation method requires that the PDEs and boundary conditions be satisfied exactly at certain points does not mean the approximate solution is exact at those points.

From (B.5) it is clear why the basis functions Φ_i for the collocation method must have continuous first derivatives: Otherwise the second derivatives that appear there will be infinite in places. For the Galerkin method (B.4) only first derivatives appear in the calculation of A_{ki}, b_k.

B.2 1D Steady-state Collocation Problems

For 1D problems, the PDE2D collocation finite element method uses cubic Hermite basis functions, that is, each unknown function $U(x)$ is approximated by

$$U(x) \approx \sum_{i=1}^{NXGRID} a_i H_i(x) + \sum_{i=1}^{NXGRID} b_i S_i(x)$$

where the basis functions H_i and S_i are defined by

$$
\begin{aligned}
H_i(x) &= 3 \left[\frac{x - x_{i-1}}{x_i - x_{i-1}} \right]^2 - 2 \left[\frac{x - x_{i-1}}{x_i - x_{i-1}} \right]^3, && \text{for} \quad x_{i-1} \leq x \leq x_i, \\
&= 3 \left[\frac{x_{i+1} - x}{x_{i+1} - x_i} \right]^2 - 2 \left[\frac{x_{i+1} - x}{x_{i+1} - x_i} \right]^3, && \text{for} \quad x_i < x \leq x_{i+1}, \\
&= 0, && \text{elsewhere,}
\end{aligned}
\tag{B.6}
$$

$$
\begin{aligned}
S_i(x) &= -\frac{(x - x_{i-1})^2}{(x_i - x_{i-1})} + \frac{(x - x_{i-1})^3}{(x_i - x_{i-1})^2}, && \text{for} \quad x_{i-1} \leq x \leq x_i, \\
&= \frac{(x_{i+1} - x)^2}{(x_{i+1} - x_i)} - \frac{(x_{i+1} - x)^3}{(x_{i+1} - x_i)^2}, && \text{for} \quad x_i < x \leq x_{i+1}, \\
&= 0, && \text{elsewhere,}
\end{aligned}
$$

These functions are piecewise cubic polynomials and are constructed so that they and their first derivatives are continuous and the collocation method, which requires continuous derivatives, can be used. It is also easy to verify directly that

$$H_i(x_j) = \delta_{ij}, \quad H'_i(x_j) = 0,$$
$$S_i(x_j) = 0, \quad S'_i(x_j) = \delta_{ij},$$

so that a_j, b_j approximate $U(x_j), U'(x_j)$.

The approximate solution is required to satisfy the PDEs exactly at two collocation points $x_j + \beta_{1,2}(x_{j+1} - x_j)$ $(\beta_1 = 0.5 - 0.5/\sqrt{3}, \beta_2 = 0.5 + 0.5/\sqrt{3})$ in each of the NXGRID-1 subintervals $(x_j, x_{j+1}), j = 1, NXGRID - 1$, and to satisfy the boundary conditions at x_1 and x_{NXGRID}. Problem 10b of chapter 5 of Sewell (2015) shows why choosing the collocation points to be these Gaussian integration points is optimal. Thus the number of boundary collocation points, 2, plus the number of interior collocation points, 2(NXGRID-1), is equal to the number of basis functions, $M = 2 \cdot NXGRID$. If there are NEQN PDEs, this ensures that the number of algebraic equations equals the number of unknowns, $N = M \cdot NEQN = 2 \cdot NXGRID \cdot NEQN$.

Newton's method, with finite difference-calculated (or user-supplied) Jacobian matrix elements, is used to solve the nonlinear algebraic equations produced when the collocation method is applied to a nonlinear steady-state PDE system, and the linear system that must be solved in each Newton iteration is solved using the Harwell Library sparse direct solver MA37, which employs a minimal degree algorithm (Duff and Reid 1984). Normally, this linear system will involve a nonsymmetric band matrix of half-bandwidth L=3*NEQN-1, which could be solved efficiently by a band solver, but when periodic boundary conditions are requested, the matrix has nonzero elements in the upper right and lower left corners, and the half-bandwidth becomes maximal. Though it is fairly easy (see problem 3 of chapter 4 of Sewell (2015)) to modify a band solver to handle such systems efficiently, a sparse direct solver such as MA37 can also handle this sparse structure efficiently. (MA37 is also used when 1D problems are solved using the Galerkin method, unless the problem is symmetric, in which case MA27 is called.) If the PDE system is linear, Newton's method is still used, but then only one iteration is necessary.

The use of cubic Hermite basis functions produces solutions with $O(h^4)$ accuracy, where $h = \max(x_{i+1} - x_i)$, which compares favorably with the $O(h^2)$ accuracy obtained when centered finite difference methods or finite element methods with linear elements are used.

If the user states (using NONE) that there is no boundary condition at a boundary point, the PDE is enforced at a point very near the boundary. In other words, the boundary collocation point is treated as an additional interior collocation point.

B.3   2D Steady-state Galerkin Problems

For 2D problems, the PDE2D Galerkin finite element method uses piecewise polynomial basis functions of degree 1, 2, 3, or 4 on triangular elements. A detailed description of the elements used is given in section 2.2 of Sewell (1985). Lagrangian finite elements are used, which means that every basis function is a piecewise polynomial that is one at "its" node and zero at every other node. Elements of degree 1, 2, 3, and 4 have 3, 6, 10, and 15 nodes, respectively, in each triangle. Thus each unknown function $U(x, y)$ is approximated by

$$U(x,y) \approx \sum_{i=1}^{M} a_i \Phi_i(x,y)$$

where, if (x_j, y_j) is a node, $\Phi_i(x_j, y_j) = \delta_{ij}$, and so a_j approximates $U(x_j, y_j)$.

The basis functions $\Phi_i(x, y)$ are piecewise polynomials that are continuous but do not have continuous first derivatives; but the Galerkin method does not require continuous first derivatives when solving second-order PDEs.

The PDE2D Galerkin method used for 1D problems also uses Lagrangian elements of degree n=1, 2, 3, or 4. The $n + 1$ nodes are chosen to coincide with the integration points, which are chosen to be optimal given that the first and last must be endpoints of the interval.

Figure B.1 shows the locations of the nodes for each element and of the integration points used (when IDEG> 0) in the computation of the integrals required by the Galerkin method. It is known (section 5.3 of Sewell (2015)) that the integration rules must be exact for polynomials of degree $2n - 2$ in order to preserve the $O(h^{n+1})$ accuracy normally expected for elements of degree n; these rules are exact for polynomials of degree 2, 5, 6, and 8, when n=1, 2, 3, and 4, and thus are more than accurate enough. The integration rules used to compute the boundary integrals use the boundary nodes as integration points and are exact for polynomials of degrees 1, 3, 5, and 7, respectively.

The "isoparametric" method is used in triangles adjacent to a curved boundary; this means that the nodes on one edge of the triangle are moved to interpolate the boundary and a change of coordinates is used to map the nodes of the original (straight) triangle onto the nodes of the curved triangle. The change of coordinates used involves polynomials of the same degree as the basis functions. This ensures that the accuracy of the element is still $O(h^{n+1})$ when the boundary is curved. In fact, since PDE2D uses isoparametric elements along a curved interface (e.g. Example 3.2), it produces optimal order accuracy even when there are curved interfaces.

Again, Newton's method is used to solve the nonlinear algebraic equations resulting from the Galerkin method formulation, and there are several options available to solve the linear system, which is assembled "element by element" (see section 5.6 of Sewell (2015)), in each Newton iteration. Section I.4 includes

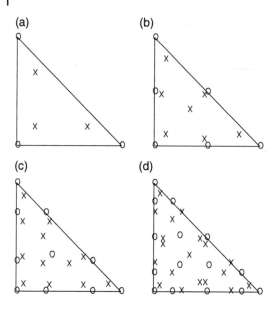

Figure B.1 PDE2D's triangular elements: (a) linear, (b) quadratic, (c) cubic, and (d) quartic. O = nodes, X = integration points.

a list and short description of these linear system solvers, with results from some tests.

The PDE2D user normally specifies an initial triangulation "by hand" (e.g. Figure I.4), with just enough triangles to define the region (usually 3–10 triangles, for simple regions). Curved boundaries are specified either by their parametric equations or by a set of boundary points, through which PDE2D draws a cubic spline. If there are points on a boundary piece where the curved arc should not be smooth (there are several such points on the boundary of Venezuela in Figure I.1, for example), you can simply duplicate a boundary point, and continuity of the spline derivatives will not be enforced at that point.

When the region is a rectangle or a disk, or any other region that can be specified in the form $X = X(P, Q)$, $Y = Y(P, Q)$, with constant limits on P and Q, the user can alternatively specify a set of P and Q grid lines, and a (regular) initial triangulation will be generated automatically (e.g. Figure I.5), with four triangles in each (possibly curved) grid rectangle. In either case, the initial triangulation can then be refined automatically until the number of triangles reaches a number specified by the user. By default, the triangles will be approximately of equal size, but a graded triangulation can be generated in one of two ways: The user can specify a positive function of x and y, which is largest where he/she wants the triangulation to be most dense, or the triangulation grading can be done adaptively. If adaptive grading of the triangulation is requested, the problem is solved once with a (uniform or user-graded) triangulation, and information about the gradient of the solution is saved in a file. The problem is then resolved; this time the gradient information is read from the file, and the triangulation is refined most where the gradients are largest. The first version, 2DEPEP, of this program had one of the first adaptive triangulation schemes (Sewell 1976), which was designed to make the triangulation most dense where the $(n + 1)$st derivatives of the solution were largest (n = element degree), given (see A.4.1 of Sewell (2015)) that the error depends on these derivatives. However, this algorithm was abandoned in a few years, because of the unreliability of the estimates of the $(n + 1)$st derivatives (after all, the $(n + 1)$st derivatives of a piecewise nth degree polynomial are zero everywhere except at the triangle boundaries, where they are infinite!). The current algorithm used by PDE2D, based on estimates of the first derivatives (regardless of the element degree), is less justifiable theoretically, but in practice it works much better than the old scheme, because estimates of the first derivatives are much more reliable, and in most problems, the size of the gradients is a good indicator of where the triangulation needs to be most dense. In any case, if the user does not like the adaptively generated triangulation, he/she can control it manually.

B.4 3D Steady-state Collocation Problems

For three-dimensional (3D) problems, PDE2D uses a collocation finite element method, with "tricubic Hermite" basis functions. That is, each unknown is approximated by a linear combination of the $8 \cdot NXGRID \cdot NYGRID \cdot NZGRID$ basis functions:

$$H_i(x)H_j(y)H_k(z), \quad H_i(x)H_j(y)S_k(z),$$
$$H_i(x)S_j(y)H_k(z), \quad H_i(x)S_j(y)S_k(z),$$
$$S_i(x)H_j(y)H_k(z), \quad S_i(x)H_j(y)S_k(z),$$
$$S_i(x)S_j(y)H_k(z), \quad S_i(x)S_j(y)S_k(z)$$

$$(i = 1, \ldots, NXGRID, j = 1, \ldots, NYGRID, k = 1, \ldots, NZGRID),$$

where the cubic Hermite basis functions H_i and S_i are defined in (B.6). This choice of basis function ensures that the first derivatives of the approximate solution are all continuous, as required by the collocation method. The approximate solution is required to satisfy the PDEs exactly at eight collocation points $(x_i + \beta_{1,2}(x_{i+1} - x_i), \ y_j + \beta_{1,2}(y_{j+1} - y_j), \ z_k + \beta_{1,2}(z_{k+1} - z_k))$, where $\beta_1 = 0.5 - 0.5/\sqrt{3}$, $\beta_2 = 0.5 + 0.5/\sqrt{3}$, in each of the $(NXGRID-1)(NYGRID-1)$ $(NZGRID-1)$ subrectangles, and to satisfy the boundary conditions at certain boundary points. The number of boundary collocation points plus the number of interior collocation points is equal to the number of basis functions ($M = 8 \cdot NXGRID \cdot NYGRID \cdot NZGRID$), so that the number of equations equals the number of unknowns ($N = M \cdot NEQN = 8 \cdot NXGRID \cdot NYGRID \cdot NZGRID \cdot NEQN$).

Most everything said about 3D collocation problems in this section applies also to 2D collocation problems, with obvious modifications.

Although it appears from the above discussion that PDE2D can only solve 3D problems in rectangular "boxes," we will see in the next section how it can easily solve problems in a wide range of simple 3D regions, such as spheres, cylinders, tori, ellipsoids, parallelepipeds, and many others.

Again, Newton's method is used to solve the nonlinear algebraic equations resulting from the collocation method formulation, and again there are several options available to solve the linear system in each Newton iteration. Section I.4 includes a list and short description of these linear system solvers, with results from some tests. The linear systems generated by Galerkin finite element methods have symmetric nonzero structures, even when the matrices themselves are nonsymmetric. The systems generated by collocation finite element methods, on the other hand, do not even have symmetric nonzero structures, and all iterative and sparse direct solvers tested on such systems perform *extremely* poorly; even MA37 does not fare well. Thus for 3D problems, PDE2D multiplies both sides of the linear system to be solved, $Ax = b$, by A^T, yielding the "normal" equations, $A^T Ax = A^T b$, and solves them using either the Harwell symmetric solver MA27 (ISOLVE=1) or a diagonal-preconditioned conjugate-gradient iterative method (ISOLVE=3). (The out-of-core and parallel band solvers, ISOLVE=2 and 6, still solve the original system $Ax = b$.) Since $A^T A$ is always symmetric and positive definite, the normal equations can be solved efficiently by many sparse direct or iterative methods. When the iterative method is used, the matrix $A^T A$ is never explicitly formed; instead, the multiplication $A^T Ap$ of a vector p is just done using two sparse matrix multiplications. $A^T A$ *is* explicitly formed when the sparse direct solver is used, but this matrix is still sparse. In fact, the normal equations are essentially the equations that would result if a "least squares" finite element method were used (see problem 10 of chapter 5 of Sewell (2015)), and the nonzero structure of $A^T A$ is the same as for a Galerkin method. However, the normal equations are significantly more ill conditioned than the original equations, and roundoff error can sometimes be a concern.

B.5 Nonrectangular 3D Regions

Most of the mathematical algorithms used by PDE2D are standard, well-known numerical methods. The way PDE2D solves problems in 3D nonrectangular regions, however, is quite original and unique (see Sewell (2010) for more detail). Again, most everything said in this section applies also to 2D collocation problems, with obvious modifications.

In 1994, PDE2D was generalized to handle 3D PDE systems. The logical extension of the 2D algorithm would have been to use a Galerkin method with (possibly isoparametric) tetrahedral elements and require the user to supply an initial "tetrahedralization" of the region, thereby facilitating the solution of problems in general 3D regions. But developing a user interface for defining general regions and boundary conditions is a *much* more difficult problem in three dimensions than in two, for reasons that are obvious and well known. The decision was made *initially*, therefore, to avoid the difficulties in handling general 3D regions and to develop software that could solve 3D PDE systems as general as those solved by the 2D algorithm, with comparable ease of use, but only in 3D boxes.

For 3D problems, then, a collocation finite element method was selected, with tricubic Hermite basis functions, because the fact that the Galerkin method is easier to apply to general regions was now not an issue, and the collocation method has some important advantages over the Galerkin brand with regard to ease of use. In particular, the user does not have to put his/her equations in the divergence form required by the Galerkin method, and so not only the PDEs but also the boundary conditions are usually more convenient to formulate. For many scientific applications, the divergence form and "natural" boundary conditions required by the Galerkin method are indeed quite natural; but general PDEs with general boundary conditions (e.g. many equations of mathematical finance; see Chapter 4) are often difficult, sometimes impossible, to express in the required format.

After the abilities to handle periodic and "no" boundary conditions were added, it became possible to solve, with high accuracy ($O(h^4)$), problems in many simple nonrectangular domains, such as spheres, cylinders, tori, pyramids, ellipsoids, and cones, by writing the PDEs in terms of an appropriate system of variables with constant limits. However, rewriting the partial differential equations in the new coordinate system was often extremely unpleasant. For example, suppose we want to solve $\nabla^2 U = 1$, with $U = 0$ on the boundary, in a torus of major radius R_0 and minor radius R_1. A "toroidal" coordinate system can be used, where

$$X = (R_0 + P3 * \cos(P2)) * \cos(P1)$$
$$Y = (R_0 + P3 * \cos(P2)) * \sin(P1)$$
$$Z = P3 * \sin(P2) \tag{B.7}$$

Here X, Y, and Z are Cartesian coordinates, $P1$ is the major (toroidal) angle, $P2$ is the minor (poloidal) angle, and $P3$ is radial distance from the torus centerline. In the new coordinate system, the region is rectangular, because the limits on $P1$, $P2$, and $P3$ are constants, and PDE2D can be used to solve this problem, with periodic boundary conditions at $P1 = 0, 2\pi$ and also at $P2 = 0, 2\pi$, "no" boundary conditions at $P3 = 0$ (because there is no boundary!), and $U = 0$ at $P3 = R_1$. To convert the Laplacian, $U_{xx} + U_{yy} + U_{zz}$, to the new coordinate system, one has to use the chain rule; for example, U_{xx}, the second derivative of U with respect to X, is ($U_i = \frac{\partial U}{\partial Pi}, U_{ij} = \frac{\partial^2 U}{\partial Pj \partial Pi}$):

$$U_{xx} = \left(U_{11}\frac{\partial P1}{\partial X} + U_{12}\frac{\partial P2}{\partial X} + U_{13}\frac{\partial P3}{\partial X} \right)\frac{\partial P1}{\partial X}$$
$$+ \left(U_{21}\frac{\partial P1}{\partial X} + U_{22}\frac{\partial P2}{\partial X} + U_{23}\frac{\partial P3}{\partial X} \right)\frac{\partial P2}{\partial X}$$
$$+ \left(U_{31}\frac{\partial P1}{\partial X} + U_{32}\frac{\partial P2}{\partial X} + U_{33}\frac{\partial P3}{\partial X} \right)\frac{\partial P3}{\partial X}$$
$$+ U_1\frac{\partial^2 P1}{\partial X^2} + U_2\frac{\partial^2 P2}{\partial X^2} + U_3\frac{\partial^2 P3}{\partial X^2}$$

The bad news is below, where just one of the terms, $\frac{\partial^2 P3}{\partial X^2}$, in this expression is displayed for the toroidal coordinate transformation, as computed by *Mathematica*. Transforming PDEs to a new coordinate system by hand can be quite a task!

$$\frac{\partial^2 P3}{\partial X^2} = -\frac{X^2(-R_0 + \sqrt{X^2 + Y^2})^2}{(X^2 + Y^2)((-R_0 + \sqrt{X^2 + Y^2})^2 + Z^2)^{3/2}}$$
$$+ \frac{X^2}{(X^2 + Y^2)\sqrt{(-R_0 + \sqrt{X^2 + Y^2})^2 + Z^2}}$$
$$- \frac{X^2(-R_0 + \sqrt{X^2 + Y^2})}{(X^2 + Y^2)^{3/2}\sqrt{(-R_0 + \sqrt{X^2 + Y^2})^2 + Z^2}}$$
$$+ \frac{-R_0 + \sqrt{X^2 + Y^2}}{\sqrt{X^2 + Y^2}\sqrt{(-R_0 + \sqrt{X^2 + Y^2})^2 + Z^2}}$$

But now all of the pain has been removed from this process, and now PDE2D will transform PDEs and boundary conditions to the new coordinate system automatically and transparently. Now the user only has to supply the global coordinate transformation equations (e.g. B.7) and can then write his/her PDEs in their usual Cartesian form. For example, the PDE above would be written simply as $U_{xx} + U_{yy} + U_{zz} = 1$; PDE2D will automatically compute $U_{xx} + U_{yy} + U_{zz}$ in terms of $U_{11}, U_{12}, \ldots$ using the chain rule and solve the problem internally in the $P1, P2, P3$ coordinate system.

If a cylindrical or spherical coordinate system is used, PDE2D automatically supplies the first and second derivatives of $P1, P2, P3$ with respect to X, Y, Z, required to apply the chain rule; the user only has to indicate that $P1, P2, P3$ represent cylindrical ($ITRANS = \pm1$) or spherical ($ITRANS = \pm2$) coordinates. If another user-specified system is used ($ITRANS = -3$), such as toroidal coordinates, PDE2D will use finite differences to compute the first and second derivatives of X, Y, Z with respect to $P1, P2, P3$; alternatively, the user can supply these analytically ($ITRANS = 3$), though this is usually not worth the effort, and thus not recommended. Then PDE2D computes the required derivatives of $P1, P2, P3$ with respect to X, Y, Z from these. For the first derivatives, the conversion uses the fact that the Jacobian matrices for the forward and inverse transforms are inverses:

$$
J \equiv \begin{bmatrix} \frac{\partial P1}{\partial X} & \frac{\partial P1}{\partial Y} & \frac{\partial P1}{\partial Z} \\[2mm] \frac{\partial P2}{\partial X} & \frac{\partial P2}{\partial Y} & \frac{\partial P2}{\partial Z} \\[2mm] \frac{\partial P3}{\partial X} & \frac{\partial P3}{\partial Y} & \frac{\partial P3}{\partial Z} \end{bmatrix} = \begin{bmatrix} \frac{\partial X}{\partial P1} & \frac{\partial X}{\partial P2} & \frac{\partial X}{\partial P3} \\[2mm] \frac{\partial Y}{\partial P1} & \frac{\partial Y}{\partial P2} & \frac{\partial Y}{\partial P3} \\[2mm] \frac{\partial Z}{\partial P1} & \frac{\partial Z}{\partial P2} & \frac{\partial Z}{\partial P3} \end{bmatrix}^{-1}
$$

The second derivatives are calculated using ($Pi = P1, P2, P3$):

$$
Pi_H = -J^T \left[\frac{\partial Pi}{\partial X} X_H + \frac{\partial Pi}{\partial Y} Y_H + \frac{\partial Pi}{\partial Z} Z_H \right] J
$$

This formula is derived directly from the chain rule. Here the subscript H denotes the Hessian matrix of second derivatives, for example:

$$
P1_H \equiv \begin{bmatrix} \frac{\partial^2 P1}{\partial X^2} & \frac{\partial^2 P1}{\partial Y \partial X} & \frac{\partial^2 P1}{\partial Z \partial X} \\[2mm] \frac{\partial^2 P1}{\partial X \partial Y} & \frac{\partial^2 P1}{\partial Y^2} & \frac{\partial^2 P1}{\partial Z \partial Y} \\[2mm] \frac{\partial^2 P1}{\partial X \partial Z} & \frac{\partial^2 P1}{\partial Y \partial Z} & \frac{\partial^2 P1}{\partial Z^2} \end{bmatrix}
$$

Implementing the coordinate transformation did not involve any internal modifications to the PDE2D library routines, only to the function subprograms where the PDE coefficients and boundary condition coefficients are defined by the user. These functions are called by PDE2D with various values of $P1, P2, P3, U, U_1, U_2, U_3, U_{11}, \ldots$; all that had to be done was to insert code to compute $X, Y, Z, U_x, U_y, U_z, U_{xx}, \ldots$ for given $P1, P2, P3, U_1, U_2, U_3, U_{11}, \ldots$, using the chain rule. Then the user can simply define his/her PDE and boundary condition coefficients in terms of $X, Y, Z, U, U_x, U_y, U_z, U_{xx}, \ldots$, though he/she can still use the non-Cartesian variables and derivatives as well, if desired.

Consider Example 5.3, where an elasticity problem was solved in the 3D arch shown in Figure 5.3. The user only has to define his/her coordinate transformation, repeated here:

$$
X = p1
$$
$$
Y = p2 * cos(p3)
$$
$$
Z = p2 * sin(p3) \tag{B.8}
$$

The partial differential equations for the elastic body, given in Section 5.2, are repeated here for convenience:

$$AU_{xx} + BV_{yx} + BW_{zx} + C(U_{yy} + V_{xy}) + C(U_{zz} + W_{xz}) = 0$$
$$C(U_{yx} + V_{xx}) + AV_{yy} + BU_{xy} + BW_{zy} + C(V_{zz} + W_{yz}) = 0$$
$$C(U_{zx} + W_{xx}) + C(V_{zy} + W_{yy}) + AW_{zz} + BU_{xz} + BV_{yz} - 10 = 0$$

where (U, V, W) is the displacement vector and A, B, C are constants defined in Section 5.2. The free boundary conditions (applied on three of the arch faces) are also repeated here:

$$(AU_x + BV_y + BW_z)N_x + C(U_y + V_x)N_y + C(U_z + W_x)N_z = 0$$
$$C(U_y + V_x)N_x + (AV_y + BU_x + BW_z)N_y + C(V_z + W_y)N_z = 0$$
$$C(U_z + W_x)N_x + C(V_z + W_y)N_y + (AW_z + BU_x + BV_y)N_z = 0$$

Once the coordinate transformation ((B.7) or (B.8), for example) is defined, the user does not need to convert his/her PDEs or boundary conditions into the new coordinate system – saving tremendous human effort. The PDEs and boundary conditions are input almost exactly as written above, in a GUI (see Figures B.2 and B.3) or interactive driver session. Volume integrals and boundary integrals can also be written using Cartesian coordinates. The unit outward normal components N_x, N_y, N_z in Cartesian coordinates are available to the boundary condition functions and boundary integrals, as

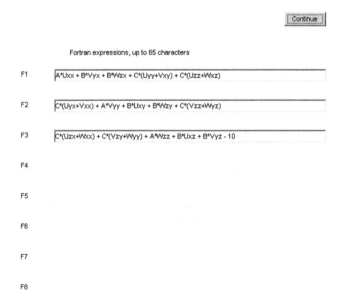

Figure B.2 GUI page defining PDEs, Example 5.3.

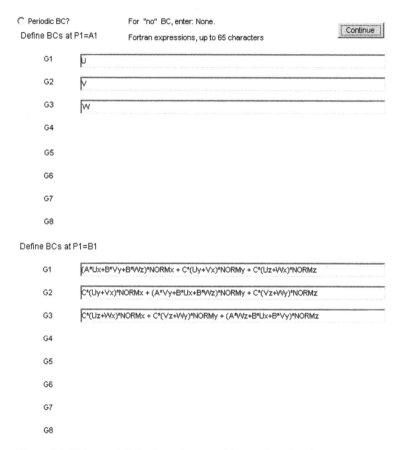

Figure B.3 GUI page defining boundary conditions at P1 = 0 and 10.

NORMx,NORMy,NORMz (see Figure B.3). In short, while PDE2D internally, hidden from the user, solves the problem in the new $P1, P2, P3$ coordinate system, the user can supply *everything* in Cartesian coordinate form.

Cross-sectional contour plots of scalar variables, and cross-sectional vector field plots, can be made that reflect the true geometry of the cross section. For example, Figure 3.15 shows the temperature field of Problem 4 of Chapter 3 at a torus cross section. For the elasticity Example 5.3, Figure 5.4 shows the displacement fields (U, V, W) at two P1=X=constant cross sections, which are plotted using axes $Y = P2 * cos(P3)$ vs. $Z = P2 * sin(P3)$, rather than $P2$ vs. $P3$, so that the cross section looks like half an annulus, as it should, rather than a rectangle.

PDE2D still cannot solve problems in complicated 3D regions, with many boundary pieces. For *simple* 3D regions, however, the coordinate transformation described here offers significant advantages over that used by other finite element method software designed to handle more general 3D regions, particularly with regard to ease of use. Once the user has supplied the transformation equations, the rest of the problem description is as simple as if the region were rectangular. Furthermore, this approach produces $O(h^4)$ accuracy even in regions with curved boundaries; without a global transformation, comparable accuracy can only be obtained if third-order *isoparametric* elements are used, and the use of high-order isoparametric elements is *much* more difficult than the global transformation approach.

B.6 Time-dependent Problems

Time-dependent problems are solved in much the same way, regardless of the spatial dimension or the finite element method (Galerkin or collocation) used. In all cases, each solution component is approximated by

$$U(x, ..., t) \approx \sum_{i=1}^{M} a_i(t)\Phi_i(x, ...), \tag{B.9}$$

where $\{\Phi_1, ..., \Phi_M\}$ is the same set of basis functions as used for the steady-state problem.

As seen in (B.9), the coefficients of the basis functions are now taken to be functions of time. When the Galerkin method is used, the expansion (B.9) is plugged into the PDEs, multiplied by a basis function Φ_k, and integrated. When the collocation method is used, the expansion (B.9) is plugged into the PDEs and required to satisfy them exactly at the collocation points. In either case, an ordinary differential equation (ODE) system results for the coefficients $a_i(t)$ in the expansion. The initial values $a_i(t_0)$ are found by interpolating the initial conditions.

For the Galerkin method, the $a_i(t_0)$ represent the values of the initial conditions $U0 \equiv U(x, ..., t_0)$ at the nodes. For the collocation method, they represent the values and derivatives of the initial conditions at the nodes, so some derivatives of the initial conditions must be interpolated. (For the 3D problem, we must interpolate $U0, U0_x, U0_y, U0_z, U0_{xy}, U0_{xz}, U0_{yz}, U0_{xyz}$.) When these derivatives are discontinuous or ill behaved, this can produce noise in the approximate (cubic Hermite) initial conditions. Thus for 1D, 2D, and 3D collocation problems, PDE2D provides an alternative: If LSQFIT=.TRUE. the cubic Hermite approximation will interpolate the values only of $U0$ at the collocation points. This is equivalent to a least squares fit to the initial values, and that tends to diminish noise in the initial conditions. See Problem 5b of

Chapter 3 for a problem where LSQFIT dramatically reduces noise. However, if LSQFIT=.TRUE., one extra linear system solution is required.

This (usually stiff) ODE system is solved using either the first-order backward difference, or backward Euler, method or the second-order Adams–Moulton method, also known in this context as the Crank–Nicolson method. Both methods are implicit, so the solution of a large (generally nonlinear) system is required each time step. Newton's method is used to solve this nonlinear system, but only one iteration is done, because a very good initial guess is available – the solution on the previous step. The linear system that must be solved each step has the same nonzero structure as in the steady-state case, and the same linear system solvers are available.

The user can either specify a constant, user-chosen step size dt or request adaptive step size control. If adaptive step size control is chosen, then each time step, two steps of size $dt/2$ are taken, and that solution is compared with the result when one step of size dt is taken. If the difference between the two answers is less than a user-supplied tolerance (for each variable), the time step dt is accepted (and the next step dt is doubled, if the agreement is *too* good); otherwise dt is halved and the process is repeated. When the step is accepted, an extrapolation is done using the answer obtained using one step of size $dt(U_1(t_{n+1}))$ and two steps of size $\frac{dt}{2}(U_2(t_{n+1}))$. When the backward Euler method is used to compute both answers, the extrapolated value is $U(t_{n+1}) = 2U_2(t_{n+1}) - U_1(t_{n+1})$, which increases the order of the truncation error of the backward Euler method from $O(dt)$ to $O(dt^2)$ while preserving its stability on stiff systems. When the Crank–Nicolson method is used, the extrapolated value is $U(t_{n+1}) = \frac{1}{2}U_2(t_{n+1}) + \frac{1}{2}U_1(t_{n+1})$, which preserves the $O(dt^2)$ truncation error of the Crank–Nicolson method, while greatly improving its stability on stiff systems. The extrapolation $U(t_{n+1}) = \frac{4}{3}U_2(t_{n+1}) - \frac{1}{3}U_1(t_{n+1})$ would increase the Crank–Nicolson truncation error order, but is not used because it produces a method that is less stable on stiff ODE systems – except for the "0D" problem, where it *is* used. It is the spatial discretization that normally makes 1D, 2D, and 3D systems stiff, so this ODE system is less likely to be stiff when 0D problems are solved.

If a constant step size is chosen, and if the problem is linear and all PDE and boundary condition coefficients are independent of time (except possibly non-homogeneous terms), then the linear system that must be solved has exactly the same coefficient matrix every time step. In this case, the *LU* decomposition computed on the first time step can be used to solve each subsequent linear system much more rapidly, and this is taken advantage of by each of the PDE2D direct linear system solvers.

PDE2D has a dump/restart option (also useful for nonlinear steady-state problems) that makes it easy to stop at some value of t, adjust the grid (or even the boundary) or "constant" time step appropriately, and restart. For example, for 2D problems the grid can be made to automatically and adaptively move with the solution, by putting a DO loop around the main program to vary the initial and final time values each pass through the loop and requesting a restart with an adaptively determined grid, as done for the KPI problem in Chapter 9. Communication between PDE2D programs is also automated; for example, it is easy to take the solution of a steady-state problem and use it as initial values for a time-dependent problem. PDE2D has functions (D)OLDSOLn, (n=1, 2, 3 for 1D, 2D, 3D problems) that allow the user to interpolate the solution saved on the last time step or iteration, at arbitrary points. This is useful, for example, if the PDE or boundary condition coefficients include an integral of the solution. Problem 4 of Chapter 4 uses DOLDSOL1 to solve an integrodifferential equation.

B.7 Eigenvalue Problems

Eigenvalue problems are also handled in much the same way regardless of the dimension of the problem or the finite element method used; only the basis functions are different. In all cases, the eigenfunction components are approximated by

$$U(x, ...) \approx \sum_{i=1}^{M} a_i \Phi_i(x, ...), \tag{B.10}$$

where $\{\Phi_1, ..., \Phi_M\}$ is the same set of basis functions as used for the steady-state problem.

When the Galerkin method is used, (B.10) is plugged into the eigenvalue PDEs, multiplied by a basis function Φ_k, and integrated. When the collocation method is used, the expansion (B.10) is plugged into the eigenvalue PDEs and is required to satisfy them exactly at the collocation points. In either case, a generalized matrix eigenvalue problem of the form $A\mathbf{a} = \lambda B\mathbf{a}$ results, for the coefficients a_i. In every case, the generalized matrix eigenvalue problem is solved using the shifted inverse power method (equation 4.11.6 of Sewell (2015)), which finds the eigenvalue closest to a user-chosen number, α (variable EV0R in the PDE2D program), and the corresponding eigenfunction. Each iteration, $(A - \alpha B)\mathbf{a}_{n+1} = B\mathbf{a}_n$, of the inverse power method requires the solution of a linear system that has the same nonzero structure as in the steady-state and time-dependent cases and can be solved using the same options. The eigenvector is renormalized each iteration to prevent overflow or underflow. The coefficient matrix $(A - \alpha B)$ of the linear system does not change each iteration, and all the direct linear system solvers used by PDE2D take advantage of this fact, forming the LU decomposition the first iteration and using it on subsequent iterations. Though the user can specify the initial values for the shifted inverse power iteration, by default they are chosen using a random number generator, which virtually eliminates the slim possibility of an "unlucky" initial vector choice (as discussed in section 4.11 of Sewell (2015)).

For 0D, 1D, 2D, and 3D eigenvalue problems, PDE2D also offers the option to compute all eigenvalues, including complex eigenvalues. The generalized discrete eigenvalue problem $A\mathbf{a} = \lambda B\mathbf{a}$ is converted to a standard eigenvalue problem $(A - qB)^{-1}B\mathbf{a} = \frac{1}{\lambda-q}\mathbf{a}$, where q (variable P8Z in the PDE2D program) is not an eigenvalue, so $(A - qB)^{-1}$ exists. Although the matrix $F \equiv (A - qB)^{-1}B$ is full, the process of finding its eigenvalues is parallelized, as follows. First, the columns f_i of the full matrix F are found by solving $(A - qB)f_i = b_i$, where b_i is the ith column of B. Since $A - qB$ is banded, the first column is found using a band solver, saving the banded LU decomposition, which is then used to solve for the other columns, with the different f_i computed on different processors if a multiprocessor machine is used. Next, the full matrix F is reduced to a similar upper Hessenberg matrix, using a Householder orthogonal reduction routine similar to HESSH shown in figure 3.4.2 of Sewell (2014). On multiprocessor machines, a parallelized version based on the parallel routine PHESSH shown on page 311 of Sewell (2014) is used. Finally, the eigenvalues, including complex eigenvalues, of this upper Hessenberg matrix are found using the EISPACK (Smith et al. 1974) routine HQR, which employs the shifted QR algorithm, discussed in section 3.3 of Sewell (2014). This stage is not parallelized. Of course once the eigenvalues μ_i of $F = (A - qB)^{-1}B$ are found, the eigenvalues of the original problem $A\mathbf{a} = \lambda B\mathbf{a}$ are $\lambda_i = q + 1/\mu_i$.

If A is symmetric and B is positive definite, so that it has a Cholesky decomposition $B = LL^T$ (L is lower triangular), then $A\mathbf{a} = \lambda B\mathbf{a}$ can be written as $L^{-1}AL^{-T}\mathbf{b} = \lambda \mathbf{b}$, where $\mathbf{b} = L^T\mathbf{a}$, and the matrix $H = L^{-1}AL^{-T}$ will be symmetric. H will generally be full even when A and B are banded, but if B is positive and *diagonal*, $L = L^T = B^{\frac{1}{2}}$ and $H = B^{-\frac{1}{2}}AB^{-\frac{1}{2}}$ will be symmetric and banded. Now the EISPACK routine BANDR can be used, which takes advantage of the band structure of this matrix, resulting in a dramatic savings in computer time and memory. For the collocation method, unfortunately, A is never symmetric, so this savings is only possible for 1D and 2D symmetric problems using the Galerkin method. Further, the requirement that B must be diagonal means that if $NEQN > 1$, the PDE2D "RHO" matrix must be diagonal and that the integration points must be the same as the nodes, which is the case for all 1D Galerkin elements, but for 2D Galerkin elements only if $IDEG = -1$ or -3 (for triangular elements of degrees 2 and 4, if the nodes are used as integration points, some weights will be negative or zero, and B will be diagonal but not positive). For such problems, however, the decrease in computer time is spectacular: See Tables I.6 and I.7 for results that demonstrate this.

B.8   The PDE2D Parallel Solvers

Even the sparse direct and iterative linear system solvers employed by PDE2D require prohibitively large amounts of computer time when very large 3D problems are solved, so PDE2D provides an additional option to solve linear systems efficiently on multiprocessor machines. In theory, such machines can achieve extremely high computational rates, but taking advantage of the parallel processing abilities of such computers generally requires substantial programming effort. A multiprocessor system can be thought of as consisting of several autonomous computers, each with its own memory (distributed memory system), or at least its own section of memory (shared memory systems), with the ability to pass data back and forth between computers. Although the communication speed between these "computers" (processors) is high, it is very slow compared with the speed with which the data are processed internally within a processor, so it is absolutely critical to minimize the amount of "message passing" between processors. A computation that is ideally suited for such a multiprocessor machine would be one where each processor does hours of calculations and computes one number each and then, at the end, the processors simply add their results together, like the Monte Carlo Example 4.2. Unfortunately, most algorithms require more communication between processors.

PDE2D uses MPI library routines to pass messages, and since this is a widely available library, this ensures that PDE2D can be run on many different distributed and shared memory multiprocessor systems.

PDE2D provides an MPI-based parallel band solver (ISOLVE=6), which can be used to solve the linear systems generated by 2D or 3D problems. This band solver is similar to other band solvers; the primary difference is that the columns of the band matrix are distributed cyclically over the available processors. That is, column 1 is stored only in the first processor's memory, column 2 is stored in the second processor's memory, and so on until we run out of processors, and then the next column is stored again in the first processor's memory. Of course, as with any band solver, only the portion of a column that is nonzero or may fill in during elimination is actually stored in memory; the elements outside this band are understood to be zero and never referenced.

Figure B.4 illustrates how the forward elimination proceeds when the matrix is distributed over the processors in this way, for a linear system with $N = 24$ unknowns, with a half-bandwidth of $L = 7$, when we have $NPES = 3$ processors. After the first 4 columns have been zeroed, the "active" column is column 5, held in processor 2's memory. Now we need to switch row 5 with the row corresponding to the largest element in the active column and then take a multiple $(-A_{65}/A_{55})$ of the fifth row and add it to the sixth row, another multiple $(-A_{75}/A_{55})$ of the fifth row and add it to the seventh row, and so on. Processor 3, for example, can do its share of these row operations, to "its" columns 6, 9,

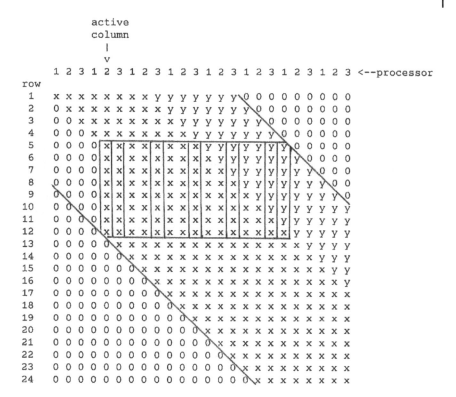

Figure B.4 Distributed band matrix. x = may be nonzero in original matrix and y = may become nonzero after pivoting.

12, 15, and 18, but it has to see the active column before it can know which row to switch with row 5 and what multiples of row 5 to add to rows 6 through 12. So processor 2 has to "broadcast" the active column to the other processors. Subroutine PLINEQ displayed in figure 6.2.1 of Sewell (2014) implements this distributed Gaussian elimination algorithm, but on full matrices. Subroutine PBAND, the "local" solver displayed in Section D.2, implements the algorithm on band matrices.

Section I.4 includes some test results for this parallel band solver on 2D and 3D problems, which show that it is primarily useful for 3D problems; but of course 3D problems are where computation speed is most critical.

When the band matrix is large, each processor has many columns to process and only one to receive (from whomever has the active column); thus the amount of communication is small compared with the amount of work done between communications. Since, for large problems, the process of adding multiples of one row to others consumes nearly 100% of the computer time when

Gaussian elimination is done, distributing this process over the available processors, allowing all processors to work simultaneously their parts of the task, ensures that the overall computation time is greatly diminished, compared with doing the whole band elimination on one processor. In fact, when the bandwidth is large compared to the number of processing elements (NPES), each processor will have approximately the same number of columns to work on, and the total work should be decreased by approximately a factor of NPES (in theory, at least!). Notice that the amount of memory required per processor is also decreased by a factor of about NPES, because the matrix is distributed nearly evenly over all the processors' memories. After the forward elimination, the solution is found by back substitution; this step is also parallelized.

The 2D and 3D iterative solvers (ISOLVE=3) have also been "MPI enhanced," that is, they also distribute the nonzeros of the matrix over the available processors. The most time-consuming calculation for these conjugate-gradient type iterative methods involves repeatedly multiplying a sparse matrix times a vector. This computation is relatively easy to distribute over the processors, each multiplying the vector by its part of the matrix. Then the different portions of this matrix–vector product, computed on the different processors, are summed together at the end, that is, $A\mathbf{p} = A_1\mathbf{p} + \cdots + A_{\text{NPES}}\mathbf{p}$, where A_i is the portion of matrix A that is stored on processor i. These solvers also distribute the other conjugate-gradient vector operations over the available processors. The MPI-enhanced routine used for 3D problems is similar to the subroutine exhibited in figure 6.2.4 of Sewell (2014), and to the "local" solver PCG listed in Section D.2. Some test results for these iterative methods are given in Section I.4.

It should be added that not only are the linear system solvers ISOLVE=3,6 parallelized, the matrix assembly is also distributed over the available processors.

Appendix C

Equations Solved by PDE2D

C.1 0D Problems

C.1.1 Steady-state Problems

$$F_1(U1, ..., UN) = 0$$
$$. \quad = .$$
$$. \quad = .$$
$$F_N(U1, ..., UN) = 0$$

C.1.2 Time-dependent Problems

$$C_{11}(t, U1, ..., UN)\frac{dU1}{dt} + \cdots + C_{1N}(t, U1, ..., UN)\frac{dUN}{dt} =$$
$$F_1(t, U1, ..., UN)$$

$$. \quad =$$
$$. \quad =$$

$$C_{N1}(t, U1, ..., UN)\frac{dU1}{dt} + \cdots + C_{NN}(t, U1, ..., UN)\frac{dUN}{dt} =$$
$$F_N(t, U1, ..., UN)$$

Initial conditions:

$$U1(t_0) = U1_0$$
$$. \quad = .$$
$$. \quad = .$$
$$UN(t_0) = UN_0$$

Solving Partial Differential Equation Applications with PDE2D, First Edition. Granville Sewell.
© 2018 John Wiley & Sons, Inc. Published 2018 by John Wiley & Sons, Inc.

C.1.3 Eigenvalue Problems

(Equations must be linear.)

$$F_1(U1, ..., UN) = \lambda \rho_{11} U1 + \cdots + \lambda \rho_{1N} UN$$

$$. \quad = \quad .$$

$$. \quad = \quad .$$

$$F_N(U1, ..., UN) = \lambda \rho_{N1} U1 + \cdots + \lambda \rho_{NN} UN$$

C.2   1D Problems (Galerkin Method)

C.2.1   Steady-state Problems

$$\frac{\partial}{\partial x} A_1(x, U1, U1_x, ..., UN, UN_x) = F_1(x, U1, U1_x, ..., UN, UN_x)$$

$$. \quad = \quad .$$

$$. \quad = \quad .$$

$$\frac{\partial}{\partial x} A_N(x, U1, U1_x, ..., UN, UN_x) = F_N(x, U1, U1_x, ..., UN, UN_x)$$

Boundary conditions (at endpoints):

$$U1 = FB_1$$

$$. \quad = \quad .$$

$$. \quad = \quad .$$

$$UN = FB_N$$

or ($N_x = -1$ at left end, $+1$ at right end)

$$A_1 N_x = GB_1(U1, U1_x, ..., UN, UN_x)$$

$$. \quad = \quad .$$

$$. \quad = \quad .$$

$$A_N N_x = GB_N(U1, U1_x, ..., UN, UN_x)$$

C.2.2   Time-dependent Problems

$$C_{11} \frac{\partial U1}{\partial t} + \cdots + C_{1N} \frac{\partial UN}{\partial t} = \frac{\partial}{\partial x} A_1(x, t, U1, U1_x, ..., UN, UN_x)$$
$$- F_1(x, t, U1, U1_x, ..., UN, UN_x)$$

$$. \quad = \quad .$$

$$. \quad = \quad .$$

$$C_{N1} \frac{\partial U1}{\partial t} + \cdots + C_{NN} \frac{\partial UN}{\partial t} = \frac{\partial}{\partial x} A_N(x, t, U1, U1_x, ..., UN, UN_x)$$
$$- F_N(x, t, U1, U1_x, ..., UN, UN_x)$$

where the C_{ij} are functions of $(x, t, U1, ..., UN)$.

Boundary conditions (at endpoints):

$$U1 = FB_1(t)$$

$$. = .$$

$$. = .$$

$$UN = FB_N(t)$$

or ($N_x = -1$ at left end, $+1$ at right end)

$$A_1 N_x = GB_1(t, U1, U1_x, ..., UN, UN_x)$$

$$. = .$$

$$. = .$$

$$A_N N_x = GB_N(t, U1, U1_x, ..., UN, UN_x)$$

Initial conditions:

$$U1(x, t_0) = U1_0(x)$$

$$. = .$$

$$. = .$$

$$UN(x, t_0) = UN_0(x)$$

C.2.3 Eigenvalue Problems

(Equations must be linear.)

$$\frac{\partial}{\partial x} A_1(x, U1, U1_x, ..., UN, UN_x) = F_1(x, U1, U1_x, ..., UN, UN_x)$$
$$+ \lambda \rho_{11}(x)U1 + \cdots + \lambda \rho_{1N}(x)UN$$

$$. = .$$

$$. = .$$

$$\frac{\partial}{\partial x} A_N(x, U1, U1_x, ..., UN, UN_x) = F_N(x, U1, U1_x, ..., UN, UN_x)$$
$$+ \lambda \rho_{N1}(x)U1 + \cdots + \lambda \rho_{NN}(x)UN$$

Boundary conditions (at endpoints):

$$U1 = FB_1$$

$$. = .$$

$$. = .$$

$$UN = FB_N$$

or ($N_x = -1$ at left end, $+1$ at right end)

$$A_1 N_x = GB_1(U1, U1_x, ..., UN, UN_x)$$

$$. \quad = \quad .$$

$$. \quad = \quad .$$

$$A_N N_x = GB_N(U1, U1_x, ..., UN, UN_x)$$

C.3 2D Problems (Galerkin Method)

C.3.1 Steady-state Problems

$$\frac{\partial}{\partial x} A_1(x, y, U1, U1_x, U1_y, ..., UN, UN_x, UN_y)$$

$$+ \frac{\partial}{\partial y} B_1(x, y, U1, U1_x, U1_y, ..., UN, UN_x, UN_y) =$$

$$F_1(x, y, U1, U1_x, U1_y, ..., UN, UN_x, UN_y)$$

$$. \quad = \quad$$

$$. \quad = \quad$$

$$\frac{\partial}{\partial x} A_N(x, y, U1, U1_x, U1_y, ..., UN, UN_x, UN_y)$$

$$+ \frac{\partial}{\partial y} B_N(x, y, U1, U1_x, U1_y, ..., UN, UN_x, UN_y) =$$

$$F_N(x, y, U1, U1_x, U1_y, ..., UN, UN_x, UN_y)$$

Boundary conditions:

$$U1 = FB_1(x, y)$$

$$. \quad = \quad .$$

$$. \quad = \quad .$$

$$UN = FB_N(x, y)$$

or

$$A_1 N_x + B_1 N_y = GB_1(x, y, U1, U1_x, U1_y, ..., UN, UN_x, UN_y)$$

$$. \quad = \quad .$$

$$. \quad = \quad .$$

$$A_N N_x + B_N N_y = GB_N(x, y, U1, U1_x, U1_y, ..., UN, UN_x, UN_y)$$

where (N_x, N_y) = unit outward normal vector.

C.3.2 Time-dependent Problems

$$C_{11}\frac{\partial U1}{\partial t} + \cdots + C_{1N}\frac{\partial UN}{\partial t} = \frac{\partial}{\partial x}A_1(x,y,t,U1,U1_x,U1_y,...,UN,UN_x,UN_y)$$
$$+\frac{\partial}{\partial y}B_1(x,y,t,U1,U1_x,U1_y,...,UN,UN_x,UN_y)$$
$$-F_1(x,y,t,U1,U1_x,U1_y,...,UN,UN_x,UN_y)$$

$$\cdot \quad = \quad \cdot$$
$$\cdot \quad = \quad \cdot$$
$$\cdot \quad = \quad \cdot$$

$$C_{N1}\frac{\partial U1}{\partial t} + \cdots + C_{NN}\frac{\partial UN}{\partial t} = \frac{\partial}{\partial x}A_N(x,y,t,U1,U1_x,U1_y,...,UN,UN_x,UN_y)$$
$$+\frac{\partial}{\partial y}B_N(x,y,t,U1,U1_x,U1_y,...,UN,UN_x,UN_y)$$
$$-F_N(x,y,t,U1,U1_x,U1_y,...,UN,UN_x,UN_y)$$

where the C_{ij} are functions of $(x,y,t,U1,...,UN)$.
Boundary conditions:

$$U1 = FB_1(x,y,t)$$
$$\cdot \quad = \quad \cdot$$
$$\cdot \quad = \quad \cdot$$
$$UN = FB_N(x,y,t)$$

or

$$A_1N_x + B_1N_y = GB_1(x,y,t,U1,U1_x,U1_y,...,UN,UN_x,UN_y)$$
$$\cdot \quad = \quad \cdot$$
$$\cdot \quad = \quad \cdot$$
$$A_NN_x + B_NN_y = GB_N(x,y,t,U1,U1_x,U1_y,...,UN,UN_x,UN_y)$$

where (N_x,N_y) = unit outward normal vector.
Initial conditions:

$$U1(x,y,t_0) = U1_0(x,y)$$
$$\cdot \quad = \quad \cdot$$
$$\cdot \quad = \quad \cdot$$
$$UN(x,y,t_0) = UN_0(x,y)$$

C.3.3 Eigenvalue Problems

(Equations must be linear.)

$$\frac{\partial}{\partial x}A_1(x,y,U1,U1_x,U1_y,...,UN,UN_x,UN_y)$$
$$+\frac{\partial}{\partial y}B_1(x,y,U1,U1_x,U1_y,...,UN,UN_x,UN_y) =$$

$$F_1(x, y, U1, U1_x, U1_y, ..., UN, UN_x, UN_y)$$
$$+ \lambda \rho_{11}(x, y)U1 + \cdots + \lambda \rho_{1N}(x, y)UN$$

$$. \quad =$$
$$. \quad =$$

$$\frac{\partial}{\partial x} A_N(x, y, U1, U1_x, U1_y, ..., UN, UN_x, UN_y)$$
$$+ \frac{\partial}{\partial y} B_N(x, y, U1, U1_x, U1_y, ..., UN, UN_x, UN_y) =$$
$$F_N(x, y, U1, U1_x, U1_y, ..., UN, UN_x, UN_y)$$
$$+ \lambda \rho_{N1}(x, y)U1 + \cdots + \lambda \rho_{NN}(x, y)UN$$

Boundary conditions:

$$U1 = FB_1(x, y)$$
$$. \quad = \quad .$$
$$. \quad = \quad .$$
$$UN = FB_N(x, y)$$

or

$$A_1 N_x + B_1 N_y = GB_1(x, y, U1, U1_x, U1_y, ..., UN, UN_x, UN_y)$$
$$. \quad = \quad .$$
$$. \quad = \quad .$$
$$A_N N_x + B_N N_y = GB_N(x, y, U1, U1_x, U1_y, ..., UN, UN_x, UN_y)$$

where (N_x, N_y) = unit outward normal vector.

C.4 1D Problems (Collocation Method)

C.4.1 Steady-state Problems

$$F_1(x, U1, U1_x, U1_{xx}, ..., UN, UN_x, UN_{xx}) = 0$$
$$. \quad = .$$
$$. \quad = .$$
$$F_N(x, U1, U1_x, U1_{xx}, ..., UN, UN_x, UN_{xx}) = 0$$

Boundary conditions (at endpoints):

$$G_1(U1, U1_x, ..., UN, UN_x) = 0$$
$$. \quad = .$$
$$. \quad = .$$
$$G_N(U1, U1_x, ..., UN, UN_x) = 0$$

(Periodic and "no" boundary conditions are also permitted.)

C.4.2 Time-dependent Problems

$$C_{11}(x, t, U1, ..., UN)\frac{\partial U1}{\partial t} + \cdots + C_{1N}(x, t, U1, ..., UN)\frac{\partial UN}{\partial t} =$$
$$F_1(x, t, U1, U1_x, U1_{xx}, ..., UN, UN_x, UN_{xx})$$

$$.\quad =$$

$$.\quad =$$

$$C_{N1}(x, t, U1, ..., UN)\frac{\partial U1}{\partial t} + \cdots + C_{NN}(x, t, U1, ..., UN)\frac{\partial UN}{\partial t} =$$
$$F_N(x, t, U1, U1_x, U1_{xx}, ..., UN, UN_x, UN_{xx})$$

Boundary conditions (at endpoints):

$$G_1(t, U1, U1_x, ..., UN, UN_x) = 0$$

$$.\quad =.$$

$$.\quad =.$$

$$G_N(t, U1, U1_x, ..., UN, UN_x) = 0$$

(Periodic and "no" boundary conditions are also permitted.)
Initial conditions:

$$U1(x, t_0) = U1_0(x)$$

$$.\quad =\quad.$$

$$.\quad =\quad.$$

$$UN(x, t_0) = UN_0(x)$$

C.4.3 Eigenvalue Problems

(Equations must be linear.)

$$F_1(x, U1, U1_x, U1_{xx}, ..., UN, UN_x, UN_{xx}) = \lambda\rho_{11}(x)U1 + \cdots + \lambda\rho_{1N}(x)UN$$

$$.\quad =\quad.$$

$$.\quad =\quad.$$

$$F_N(x, U1, U1_x, U1_{xx}, ..., UN, UN_x, UN_{xx}) = \lambda\rho_{N1}(x)U1 + \cdots + \lambda\rho_{NN}(x)UN$$

Boundary conditions (at endpoints):

$$G_1(U1, U1_x, ..., UN, UN_x) = 0$$

$$.\quad =.$$

$$.\quad =.$$

$$G_N(U1, U1_x, ..., UN, UN_x) = 0$$

(Periodic and "no" boundary conditions are also permitted.)

C.5   2D Problems (Collocation Method)

C.5.1   Steady-state Problems

$$F_1(x, y, U1, U1_x, U1_y, U1_{xx}, U1_{yy}, U1_{xy}, U2, ...) = 0$$

$$. \qquad = .$$

$$. \qquad = .$$

$$F_N(x, y, U1, U1_x, U1_y, U1_{xx}, U1_{yy}, U1_{xy}, U2, ...) = 0$$

Boundary conditions:

$$G_1(x, y, U1, U1_x, U1_y, ..., UN, UN_x, UN_y) = 0$$

$$. \qquad = .$$

$$. \qquad = .$$

$$G_N(x, y, U1, U1_x, U1_y, ..., UN, UN_x, UN_y) = 0$$

(Periodic and "no" boundary conditions are also permitted.)

C.5.2   Time-dependent Problems

$$C_{11}(x, y, t, U1, ..., UN)\frac{\partial U1}{\partial t} + \cdots + C_{1N}(x, y, t, U1, ..., UN)\frac{\partial UN}{\partial t} =$$
$$F_1(x, y, t, U1, U1_x, U1_y, U1_{xx}, U1_{yy}, U1_{xy}, U2, ...)$$

$$. \qquad =$$

$$. \qquad =$$

$$C_{N1}(x, y, t, U1, ..., UN)\frac{\partial U1}{\partial t} + \cdots + C_{NN}(x, y, t, U1, ..., UN)\frac{\partial UN}{\partial t} =$$
$$F_N(x, y, t, U1, U1_x, U1_y, U1_{xx}, U1_{yy}, U1_{xy}, U2, ...)$$

Boundary conditions:

$$G_1(x, y, t, U1, U1_x, U1_y, ..., UN, UN_x, UN_y) = 0$$

$$. \qquad = .$$

$$. \qquad = .$$

$$G_N(x, y, t, U1, U1_x, U1_y, ..., UN, UN_x, UN_y) = 0$$

(Periodic and "no" boundary conditions are also permitted.)
Initial conditions:

$$U1(x, y, t_0) = U1_0(x, y)$$

$$. \qquad = \qquad .$$

$$. \qquad = \qquad .$$

$$UN(x, y, t_0) = UN_0(x, y)$$

C.5.3 Eigenvalue Problems

(Equations must be linear.)

$$F_1(x, y, U1, U1_x, U1_y, U1_{xx}, U1_{yy}, U1_{xy}, U2, ...) =$$
$$\lambda \rho_{11}(x, y)U1 + \cdots + \lambda \rho_{1N}(x, y)UN$$
$$. \quad =$$
$$. \quad =$$
$$F_N(x, y, U1, U1_x, U1_y, U1_{xx}, U1_{yy}, U1_{xy}, U2, ...) =$$
$$\lambda \rho_{N1}(x, y)U1 + \cdots + \lambda \rho_{NN}(x, y)UN$$

Boundary conditions:

$$G_1(x, y, U1, U1_x, U1_y, ..., UN, UN_x, UN_y) = 0$$
$$. \quad = .$$
$$. \quad = .$$
$$G_N(x, y, U1, U1_x, U1_y, ..., UN, UN_x, UN_y) = 0$$

(Periodic and "no" boundary conditions are also permitted.)

C.6 3D Problems

C.6.1 Steady-state Problems

$$F_1(x, y, z, U1, U1_x, U1_y, U1_z, U1_{xx}, U1_{yy}, U1_{zz}, U1_{xy}, U1_{xz}, U1_{yz}, U2, ...) = 0$$
$$. \quad = .$$
$$. \quad = .$$
$$F_N(x, y, z, U1, U1_x, U1_y, U1_z, U1_{xx}, U1_{yy}, U1_{zz}, U1_{xy}, U1_{xz}, U1_{yz}, U2, ...) = 0$$

Boundary conditions:

$$G_1(x, y, z, U1, U1_x, U1_y, U1_z, ..., UN, UN_x, UN_y, UN_z) = 0$$
$$. \quad = .$$
$$. \quad = .$$
$$G_N(x, y, z, U1, U1_x, U1_y, U1_z, ..., UN, UN_x, UN_y, UN_z) = 0$$

(Periodic and "no" boundary conditions are also permitted.)

C.6.2 Time-dependent Problems

$$C_{11}(x,y,z,t,U1,...,UN)\frac{\partial U1}{\partial t} + \cdots + C_{1N}(x,y,z,t,U1,...,UN)\frac{\partial UN}{\partial t} =$$
$$F_1(x,y,z,t,U1,U1_x,U1_y,U1_z,U1_{xx},U1_{yy},U1_{zz},U1_{xy},U1_{xz},U1_{yz},U2,...)$$

$$. =$$
$$. =$$

$$C_{N1}(x,y,z,t,U1,...,UN)\frac{\partial U1}{\partial t} + \cdots + C_{NN}(x,y,z,t,U1,...,UN)\frac{\partial UN}{\partial t} =$$
$$F_N(x,y,z,t,U1,U1_x,U1_y,U1_z,U1_{xx},U1_{yy},U1_{zz},U1_{xy},U1_{xz},U1_{yz},U2,...)$$

Boundary conditions:

$$G_1(x,y,z,t,U1,U1_x,U1_y,U1_z,...,UN,UN_x,UN_y,UN_z) = 0$$

$$. = .$$
$$. = .$$

$$G_N(x,y,z,t,U1,U1_x,U1_y,U1_z,...,UN,UN_x,UN_y,UN_z) = 0$$

(Periodic and "no" boundary conditions are also permitted.)
Initial conditions:

$$U1(x,y,z,t_0) = U1_0(x,y,z)$$

$$. = .$$
$$. = .$$

$$UN(x,y,z,t_0) = UN_0(x,y,z)$$

C.6.3 Eigenvalue Problems

(Equations must be linear.)

$$F_1(x,y,z,U1,U1_x,U1_y,U1_z,U1_{xx},U1_{yy},U1_{zz},U1_{xy},U1_{xz},U1_{yz},U2,...) =$$
$$\lambda\rho_{11}(x,y,z)U1 + \cdots + \lambda\rho_{1N}(x,y,z)UN$$

$$. =$$
$$. =$$

$$F_N(x,y,z,U1,U1_x,U1_y,U1_z,U1_{xx},U1_{yy},U1_{zz},U1_{xy},U1_{xz},U1_{yz},U2,...) =$$
$$\lambda\rho_{N1}(x,y,z)U1 + \cdots + \lambda\rho_{NN}(x,y,z)UN$$

Boundary conditions:

$$G_1(x,y,z,U1,U1_x,U1_y,U1_z,...,UN,UN_x,UN_y,UN_z) = 0$$

$$. = .$$
$$. = .$$

$$G_N(x,y,z,U1,U1_x,U1_y,U1_z,...,UN,UN_x,UN_y,UN_z) = 0$$

(Periodic and "no" boundary conditions are also permitted.)

Appendix D

Problem 5.7 Local Solvers

D.1 DTD3M,DTD3N,DCG (problem57.f)

```
- - - - - - DTD3M - - - - - - - - - - - -

      SUBROUTINE DTD3M(N,NZ,IR,IC,A,B,JOB,SPD)
      IMPLICIT DOUBLE PRECISION (A-H,O-Z)
      DIMENSION IR(NZ),IC(NZ),A(NZ),B(N)
      LOGICAL SPD
      COMMON /DTDP27/ ITASK,NPES,ICOMM
C
C  DTD3M IS A LOCALLY-WRITTEN CODE WHICH SOLVES A SPARSE SYMMETRIC
C  LINEAR SYSTEM A*X=B.  IT WILL BE CALLED WHEN ISOLVE=5 FOR A SYMMETRIC
C  2D GALERKIN PROBLEM, AND WHEN ISOLVE=4 FOR A 2D OR 3D COLLOCATION PROBLEM.
C  IF YOU ACTIVATE DTD3M,DTD3N, YOU SHOULD ALSO INCREASE ISMX2D AND ISMX3D TO
C  AT LEAST 5 IN THE FILE 'pde2d.f'.
C
C  N  - NUMBER OF EQUATIONS AND UNKNOWNS (INPUT)
C  NZ - NUMBER OF NONZEROS IN THE UPPER TRIANGLE OF A (INPUT)
C  IR - ROW NUMBERS OF THE NONZEROS IN THE UPPER TRIANGLE OF A (INPUT)
C  IC - COLUMN NUMBERS OF THE NONZEROS IN THE UPPER TRIANGLE OF A (INPUT)
C  A  - NONZERO ELEMENTS OF THE UPPER TRIANGLE OF A (INPUT).  A(J)
C         CONTAINS ELEMENT (IR(J),IC(J)), J=1,...,NZ, OF THE MATRIX,
C         WHERE IC(J).GE.IR(J).
C  B  - ON INPUT, B WILL CONTAIN THE RIGHT HAND SIDE OF THE LINEAR
C         SYSTEM.  ON OUTPUT, B SHOULD CONTAIN THE SOLUTION, X.
C  JOB- JOB PARAMETER (INPUT).  IF JOB=2, THIS MEANS THAT THE MATRIX
C         A HAS CHANGED SINCE THE LAST CALL TO DTD3M, WHILE JOB=3 MEANS
C         A HAS NOT CHANGED.  THUS, IF YOU WISH, YOU CAN COMPUTE AN LU
C         DECOMPOSITION OF A WHEN JOB=2 AND SAVE IT, AND USE THIS
C         DECOMPOSITION TO SOLVE THE SYSTEM MORE RAPIDLY, WHEN JOB=3.
C  SPD- .TRUE. IF LINEAR SYSTEM IS POSITIVE DEFINITE. (INPUT)
C
C  IF MORE THAN ONE PROCESSOR IS USED (NPES > 1), THE MATRIX WILL BE
C  DISTRIBUTED OVER THE PROCESSORS, AND ONLY THOSE NONZEROS WITH
C  MOD(IC(J)-1,NPES) = ITASK WILL BE PASSED TO DTD3M ON PROCESSOR ITASK.
C
```

```
C       CALL DTDPS (3,
C       & ' Local solver not available.  Try another value for ISOLVE.',
C       & 0.0D0,0)
        DIMENSION X(N)
        CALL DCG(A,IR,IC,NZ,X,B,N,.TRUE.)
C            DTD3M EXPECTS THE SOLUTION TO BE RETURNED IN B.
        B = X
        RETURN
        END


- - - - - - - - DTD3N - - - - - - - - - - - - - - - - - - - -


        SUBROUTINE DTD3N(N,NZ,IR,IC,A,B,JOB)
        IMPLICIT DOUBLE PRECISION (A-H,O-Z)
        DIMENSION IR(NZ),IC(NZ),A(NZ),B(N)
        COMMON /DTDP27/ ITASK,NPES,ICOMM
C
C DTD3N IS A LOCALLY-WRITTEN CODE WHICH SOLVES A SPARSE NONSYMMETRIC
C LINEAR SYSTEM A*X=B.  IT WILL BE CALLED WHEN ISOLVE=5 FOR A NONSYMMETRIC
C 2D GALERKIN PROBLEM, AND WHEN ISOLVE=5 FOR A 2D OR 3D COLLOCATION PROBLEM.
C IF YOU ACTIVATE DTD3M,DTD3N, YOU SHOULD ALSO INCREASE ISMX2D AND ISMX3D
C TO AT LEAST 5 IN THE FILE 'pde2d.f'.
C
C N  - NUMBER OF EQUATIONS AND UNKNOWNS (INPUT)
C NZ - NUMBER OF NONZEROS IN A (INPUT)
C IR - ROW NUMBERS OF THE NONZEROS OF A (INPUT)
C IC - COLUMN NUMBERS OF THE NONZEROS OF A (INPUT)
C A  - NONZERO ELEMENTS OF A (INPUT).  A(J) CONTAINS ELEMENT
C          (IR(J),IC(J)), J=1,...,NZ, OF THE MATRIX.
C B  - ON INPUT, B WILL CONTAIN THE RIGHT HAND SIDE OF THE LINEAR
C          SYSTEM.  ON OUTPUT, B SHOULD CONTAIN THE SOLUTION, X.
C JOB- JOB PARAMETER (INPUT).  IF JOB=2, THIS MEANS THAT THE MATRIX
C          A HAS CHANGED SINCE THE LAST CALL TO DTD3N, WHILE JOB=3 MEANS
C          A HAS NOT CHANGED.  THUS, IF YOU WISH, YOU CAN COMPUTE AN LU
C          DECOMPOSITION OF A WHEN JOB=2 AND SAVE IT, AND USE THIS
C          DECOMPOSITION TO SOLVE THE SYSTEM MORE RAPIDLY, WHEN JOB=3.
C
C IF MORE THAN ONE PROCESSOR IS USED (NPES > 1), THE MATRIX WILL BE
C DISTRIBUTED OVER THE PROCESSORS, AND ONLY THOSE NONZEROS WITH
C MOD(IC(J)-1,NPES) = ITASK WILL BE PASSED TO DTD3N ON PROCESSOR ITASK.
C
C       CALL DTDPS (3,
C       & ' Local solver not available.  Try another value for ISOLVE.',
C       & 0.0D0,0)
        DIMENSION X(N)
        CALL DCG(A,IR,IC,NZ,X,B,N,.FALSE.)
C            DTD3N EXPECTS THE SOLUTION TO BE RETURNED IN B.
        B = X
        RETURN
        END


- - - - - - - - - DCG - - - - - - - - - - - - - - - - - - - -


        SUBROUTINE DCG(A,IROW,JCOL,NZ,X,B,N,SYMM)
        IMPLICIT DOUBLE PRECISION (A-H,O-Z)
```

```
C                                   DECLARATIONS FOR ARGUMENTS
      DOUBLE PRECISION A(NZ),B(N),X(N)
      INTEGER IROW(NZ),JCOL(NZ)
      LOGICAL SYMM
C                                   DECLARATIONS FOR LOCAL VARIABLES
      DOUBLE PRECISION R(N),P(N),AP(N),D(N),LAMBDA
C
C SUBROUTINE DCG SOLVES THE LINEAR SYSTEM A*X=B, USING THE JACOBI
C    CONJUGATE GRADIENT ITERATIVE METHOD.  THE NON-ZEROS OF A ARE
C    STORED IN SPARSE FORMAT.
C
C ARGUMENTS
C
C              ON INPUT                      ON OUTPUT
C              - - - -                       - - - -
C
C    A       - A(IZ) IS THE MATRIX ELEMENT IN
C              ROW IROW(IZ), COLUMN JCOL(IZ),
C              FOR IZ=1,...,NZ.
C
C    IROW    - (SEE A).
C
C    JCOL    - (SEE A).
C
C    NZ      - NUMBER OF NONZEROS.
C
C    X       -                             AN N-VECTOR CONTAINING
C                                          THE SOLUTION.
C
C    B       - THE RIGHT HAND SIDE N-VECTOR.
C
C    N       - SIZE OF MATRIX A.
C
C    SYMM    - .TRUE. IF A IS SYMMETRIC, IN
C              WHICH CASE ONLY THE NONZEROS
C              IN THE UPPER TRIANGLE ARE INPUT
C- - - - - - - - - - - - - - - - - - - - - - - - - - - - - - - - - -
C
      DO 5 I=1,NZ
C                                   DCG REQUIRES THE DIAGONAL OF A, FOR
C                                   "JACOBI" PRECONDITIONING
         IF (IROW(I).EQ.JCOL(I)) D(IROW(I)) = A(I)
    5 CONTINUE
C                                   X0 = 0
C                                   R0 = D**(-1)*B
C                                   P0 = R0
      R0MAX = 0
      DO 10 I=1,N
         X(I) = 0
         R(I) = B(I)/D(I)
         R0MAX = MAX(R0MAX,ABS(R(I)))
         P(I) = R(I)
   10 CONTINUE
C                                   NITER = MAX NUMBER OF ITERATIONS
      NITER = 3*N
```

```
      DO 90 ITER=1,NITER
C                                      AP = A*P
         DO 20 I=1,N
           AP(I) = 0
   20    CONTINUE
         DO 30 IZ=1,NZ
           I = IROW(IZ)
           J = JCOL(IZ)
           AP(I) = AP(I) + A(IZ)*P(J)
           IF (I.NE.J .AND. SYMM) AP(J) = AP(J) + A(IZ)*P(I)
   30    CONTINUE
C                                      PAP = (P,AP)
C                                      RP = (R,D*P)
         PAP = 0.0
         RP = 0.0
         DO 40 I=1,N
           PAP = PAP + P(I)*AP(I)
           RP = RP + R(I)*D(I)*P(I)
   40    CONTINUE
C                                      LAMBDA = (R,D*P)/(P,AP)
         LAMBDA = RP/PAP
C                                      X = X + LAMBDA*P
C                                      R = R - LAMBDA*D**(-1)*AP
         DO 50 I=1,N
           X(I) = X(I) + LAMBDA*P(I)
           R(I) = R(I) - LAMBDA*AP(I)/D(I)
   50    CONTINUE
C                                      RAP = (R,AP)
         RAP = 0.0
         DO 60 I=1,N
           RAP = RAP + R(I)*AP(I)
   60    CONTINUE
C                                      ALPHA = -(R,AP)/(P,AP)
         ALPHA = -RAP/PAP
C                                      P = R + ALPHA*P
         DO 70 I=1,N
           P(I) = R(I) + ALPHA*P(I)
   70    CONTINUE
C                                      RMAX = MAX OF RESIDUAL (R)
         RMAX = 0
         DO 80 I=1,N
           RMAX = MAX(RMAX,ABS(R(I)))
   80    CONTINUE
C                                 CHECK IF CONVERGED
         IF (RMAX.LE.1.D-10*R0MAX) THEN
           PRINT *, ' Number of iterations = ',ITER
           RETURN
         ENDIF
   90 CONTINUE
C                                      DCG DOES NOT CONVERGE
      PRINT 100
  100 FORMAT(' ***** DCG does not converge *****')
      RETURN
      END
```

D.2 DTD3M,DTD3N,PCG,PBAND (problem57d.f)

```
- - - - - - DTD3M - - - - - - - - - -

      SUBROUTINE DTD3M(N,NZ,IR,IC,A,B,JOB,SPD)
      IMPLICIT DOUBLE PRECISION (A-H,O-Z)
      DIMENSION IR(NZ),IC(NZ),A(NZ),B(N)
      LOGICAL SPD
      COMMON /DTDP27/ ITASK,NPES,ICOMM
C
C  DTD3M IS A LOCALLY-WRITTEN CODE WHICH SOLVES A SPARSE SYMMETRIC
C  LINEAR SYSTEM A*X=B.  IT WILL BE CALLED WHEN ISOLVE=5 FOR A SYMMETRIC
C  2D GALERKIN PROBLEM, AND WHEN ISOLVE=4 FOR A 2D OR 3D COLLOCATION PROBLEM.
C  IF YOU ACTIVATE DTD3M,DTD3N, YOU SHOULD ALSO INCREASE ISMX2D AND ISMX3D TO
C  AT LEAST 5 IN THE FILE 'pde2d.f'.
C
C  N  - NUMBER OF EQUATIONS AND UNKNOWNS (INPUT)
C  NZ - NUMBER OF NONZEROS IN THE UPPER TRIANGLE OF A (INPUT)
C  IR - ROW NUMBERS OF THE NONZEROS IN THE UPPER TRIANGLE OF A (INPUT)
C  IC - COLUMN NUMBERS OF THE NONZEROS IN THE UPPER TRIANGLE OF A  (INPUT)
C  A  - NONZERO ELEMENTS OF THE UPPER TRIANGLE OF A (INPUT).  A(J)
C        CONTAINS ELEMENT (IR(J),IC(J)), J=1,...,NZ, OF THE MATRIX,
C        WHERE IC(J).GE.IR(J).
C  B  - ON INPUT, B WILL CONTAIN THE RIGHT HAND SIDE OF THE LINEAR
C        SYSTEM.  ON OUTPUT, B SHOULD CONTAIN THE SOLUTION, X.
C  JOB- JOB PARAMETER (INPUT).  IF JOB=2, THIS MEANS THAT THE MATRIX
C        A HAS CHANGED SINCE THE LAST CALL TO DTD3M, WHILE JOB=3 MEANS
C        A HAS NOT CHANGED.  THUS, IF YOU WISH, YOU CAN COMPUTE AN LU
C        DECOMPOSITION OF A WHEN JOB=2 AND SAVE IT, AND USE THIS
C        DECOMPOSITION TO SOLVE THE SYSTEM MORE RAPIDLY, WHEN JOB=3.
C  SPD- .TRUE. IF LINEAR SYSTEM IS POSITIVE DEFINITE. (INPUT)
C
C  IF MORE THAN ONE PROCESSOR IS USED (NPES > 1), THE MATRIX WILL BE
C  DISTRIBUTED OVER THE PROCESSORS, AND ONLY THOSE NONZEROS WITH
C  MOD(IC(J)-1,NPES) = ITASK WILL BE PASSED TO DTD3M ON PROCESSOR ITASK.
C
C     CALL DTDPS (3,
C     & ' Local solver not available.  Try another value for ISOLVE.',
C     & 0.0D0,0)
      DIMENSION X(N)
      CALL PCG(A,IR,IC,NZ,X,B,N,.TRUE.)
C         DTD3M EXPECTS THE SOLUTION TO BE RETURNED IN B.
      B = X
      RETURN
      END


- - - - - - - -DTD3N - - - - - - - - - - - - - - - - - - - -

      SUBROUTINE DTD3N(N,NZ,IR,IC,A,B,JOB)
      IMPLICIT DOUBLE PRECISION (A-H,O-Z)
      DIMENSION IR(NZ),IC(NZ),A(NZ),B(N)
      COMMON /DTDP27/ ITASK,NPES,ICOMM
C
C  DTD3N IS A LOCALLY-WRITTEN CODE WHICH SOLVES A SPARSE NONSYMMETRIC
```

```
C  LINEAR SYSTEM A*X=B.  IT WILL BE CALLED WHEN ISOLVE=5 FOR A NONSYMMETRIC
C  2D GALERKIN PROBLEM, AND WHEN ISOLVE=5 FOR A 2D OR 3D COLLOCATION PROBLEM.
C  IF YOU ACTIVATE DTD3M,DTD3N, YOU SHOULD ALSO INCREASE ISMX2D AND ISMX3D
C  TO AT LEAST 5 IN THE FILE 'pde2d.f'.
C
C  N  - NUMBER OF EQUATIONS AND UNKNOWNS (INPUT)
C  NZ - NUMBER OF NONZEROS IN A (INPUT)
C  IR - ROW NUMBERS OF THE NONZEROS OF A (INPUT)
C  IC - COLUMN NUMBERS OF THE NONZEROS OF A (INPUT)
C  A  - NONZERO ELEMENTS OF A (INPUT).  A(J) CONTAINS ELEMENT
C         (IR(J),IC(J)), J=1,...,NZ, OF THE MATRIX.
C  B  - ON INPUT, B WILL CONTAIN THE RIGHT HAND SIDE OF THE LINEAR
C         SYSTEM.  ON OUTPUT, B SHOULD CONTAIN THE SOLUTION, X.
C JOB- JOB PARAMETER (INPUT).  IF JOB=2, THIS MEANS THAT THE MATRIX
C         A HAS CHANGED SINCE THE LAST CALL TO DTD3N, WHILE JOB=3 MEANS
C         A HAS NOT CHANGED.  THUS, IF YOU WISH, YOU CAN COMPUTE AN LU
C         DECOMPOSITION OF A WHEN JOB=2 AND SAVE IT, AND USE THIS
C         DECOMPOSITION TO SOLVE THE SYSTEM MORE RAPIDLY, WHEN JOB=3.
C
C  IF MORE THAN ONE PROCESSOR IS USED (NPES > 1), THE MATRIX WILL BE
C  DISTRIBUTED OVER THE PROCESSORS, AND ONLY THOSE NONZEROS WITH
C  MOD(IC(J)-1,NPES) = ITASK WILL BE PASSED TO DTD3N ON PROCESSOR ITASK.
C
C     CALL DTDPS (3,
C    & ' Local solver not available.  Try another value for ISOLVE.',
C    & 0.0D0,0)
      DIMENSION X(N)
      include 'mpif.h'
      ALLOCATABLE ABAND(:,:)
      NLDI = 0
      NUDI = 0
      DO 10 IZ=1,NZ
         NLDI =  MAX(IR(IZ)-IC(IZ),NLDI)
         NUDI =  MAX(IC(IZ)-IR(IZ),NUDI)
   10 CONTINUE
C         NLD = NUMBER OF LOWER DIAGONALS, NUD = NUMBER OF UPPER DIAGONALS
      CALL MPI_ALLREDUCE(NLDI,NLD,1,MPI_INTEGER,MPI_MAX,ICOMM,IERR)
      CALL MPI_ALLREDUCE(NUDI,NUD,1,MPI_INTEGER,MPI_MAX,ICOMM,IERR)
      ALLOCATE (ABAND(-NUD-NLD:NLD,(N-1)/NPES+1))
C         COPY LOCAL PROCESSOR'S COLUMNS OF SPARSE MATRIX A TO
C         LOCAL PROCESSOR'S COLUMNS OF BAND MATRIX ABAND
      ABAND = 0
      DO 20 IZ=1,NZ
         I = IR(IZ)
         J = IC(IZ)
         ABAND(I-J,(J-1)/NPES+1) = A(IZ)
   20 CONTINUE
      CALL PBAND(ABAND,N,NLD,NUD,X,B)
C         DTD3N EXPECTS THE SOLUTION TO BE RETURNED IN B.
      B = X
      RETURN
      END


- - - - - - - - PCG - - - - - - - - - - - - - - - - - - - - - - -

      SUBROUTINE PCG(A,IROW,JCOL,NZ,X,B,N,SYMM)
      IMPLICIT DOUBLE PRECISION (A-H,O-Z)
```

```
C                                    DECLARATIONS FOR ARGUMENTS
      DOUBLE PRECISION A(NZ),B(N),X(N)
      INTEGER IROW(NZ),JCOL(NZ)
      LOGICAL SYMM
C                                    DECLARATIONS FOR LOCAL VARIABLES
      DOUBLE PRECISION R(N),P(N),PI(N),API(N),AP(N),DI(N),D(N),LAMBDA
      include 'mpif.h'
C
C SUBROUTINE PCG SOLVES THE LINEAR SYSTEM A*X=B, USING THE JACOBI
C    CONJUGATE GRADIENT ITERATIVE METHOD.  THE NON-ZEROS OF A ARE
C    STORED IN SPARSE FORMAT.  THE COLUMNS OF A ARE DISTRIBUTED
C    CYCLICALLY OVER THE AVAILABLE PROCESSORS.
C
C ARGUMENTS
C
C              ON INPUT                        ON OUTPUT
C              - - - - -                       - - - - -
C
C    A      - A(IZ) IS THE MATRIX ELEMENT IN
C             ROW IROW(IZ), COLUMN JCOL(IZ),
C             FOR IZ=1,...,NZ. ELEMENTS WITH
C             MOD(JCOL(IZ)-1,NPES)=ITASK
C             ARE STORED ON PROCESSOR ITASK.
C
C    IROW   - (SEE A).
C
C    JCOL   - (SEE A).
C
C    NZ     - NUMBER OF NONZEROS STORED ON
C             THE LOCAL PROCESSOR.
C
C    X      -                                  AN N-VECTOR CONTAINING
C                                              THE SOLUTION.
C
C    B      - THE RIGHT HAND SIDE N-VECTOR.
C
C    N      - SIZE OF MATRIX A.
C
C    SYMM   - .TRUE. IF A IS SYMMETRIC, IN
C             WHICH CASE ONLY THE NONZEROS
C             IN THE UPPER TRIANGLE ARE INPUT
C- - - - - - - - - - - - - - - - - - - - - - - - - - - - - - - - - -
C                                    NPES = NUMBER OF PROCESSORS
      CALL MPI_COMM_SIZE (MPI_COMM_WORLD,NPES,IERR)
C                                    ITASK = MY PROCESSOR NUMBER
      CALL MPI_COMM_RANK (MPI_COMM_WORLD,ITASK,IERR)
C
      DI(1:N) = 0
      DO 5 I=1,NZ
C                                    PCG REQUIRES THE DIAGONAL OF A, FOR
C                                    "JACOBI" PRECONDITIONING
        IF (IROW(I).EQ.JCOL(I)) DI(IROW(I)) = A(I)
    5 CONTINUE
```

```
C                                 CALL MPI_ALLREDUCE TO COLLECT DIAGONAL
      CALL MPI_ALLREDUCE(DI,D,N,MPI_DOUBLE_PRECISION,
     & MPI_SUM,MPI_COMM_WORLD,IERR)
C                                 X0 = 0
C                                 R0 = D**(-1)*B
C                                 P0 = R0
      R0MAX = 0
      DO 10 I=1,N
         X(I) = 0
         R(I) = B(I)/D(I)
         R0MAX = MAX(R0MAX,ABS(R(I)))
         P(I) = R(I)
   10 CONTINUE
C                                 NITER = MAX NUMBER OF ITERATIONS
      NITER = 3*N
      DO 90 ITER=1,NITER
C                                 AP = A*P
         DO 20 I=1,N
            API(I) = 0
   20    CONTINUE
         DO 30 IZ=1,NZ
            I = IROW(IZ)
            J = JCOL(IZ)
            API(I) = API(I) + A(IZ)*P(J)
            IF (I.NE.J .AND. SYMM) API(J) = API(J) + A(IZ)*P(I)
   30    CONTINUE
C                                 MPI_ALLREDUCE COLLECTS THE VECTORS API
C                                 (API = LOCAL(A)*P) FROM ALL PROCESSORS
C                                 AND ADDS THEM TOGETHER, THEN SENDS
C                                 THE RESULT, AP, BACK TO ALL PROCESSORS.
         CALL MPI_ALLREDUCE(API,AP,N,MPI_DOUBLE_PRECISION,
     &      MPI_SUM,MPI_COMM_WORLD,IERR)
C                                 PAP = (P,AP)
C                                 RP = (R,D*P)
         PAPI = 0.0
         RPI = 0.0
         DO 40 I=ITASK+1,N,NPES
            PAPI = PAPI + P(I)*AP(I)
            RPI = RPI + R(I)*D(I)*P(I)
   40    CONTINUE
         CALL MPI_ALLREDUCE(PAPI,PAP,1,MPI_DOUBLE_PRECISION,
     &      MPI_SUM,MPI_COMM_WORLD,IERR)
         CALL MPI_ALLREDUCE(RPI,RP,1,MPI_DOUBLE_PRECISION,
     &      MPI_SUM,MPI_COMM_WORLD,IERR)
C                                 LAMBDA = (R,D*P)/(P,AP)
         LAMBDA = RP/PAP
C                                 X = X + LAMBDA*P
C                                 R = R - LAMBDA*D**(-1)*AP
         DO 50 I=ITASK+1,N,NPES
            X(I) = X(I) + LAMBDA*P(I)
            R(I) = R(I) - LAMBDA*AP(I)/D(I)
   50    CONTINUE
C                                 RAP = (R,AP)
         RAPI = 0.0
         DO 60 I=ITASK+1,N,NPES
```

```
             RAPI = RAPI + R(I)*AP(I)
      60     CONTINUE
             CALL MPI_ALLREDUCE(RAPI,RAP,1,MPI_DOUBLE_PRECISION,
      &        MPI_SUM,MPI_COMM_WORLD,IERR)
C                               ALPHA = -(R,AP)/(P,AP)
             ALPHA = -RAP/PAP
C                               P = R + ALPHA*P
                IF (SYMM) THEN
             PI(1:N) = 0
             DO 65 I=ITASK+1,N,NPES
                PI(I) = R(I) + ALPHA*P(I)
      65     CONTINUE
             CALL MPI_ALLREDUCE(PI,P,N,MPI_DOUBLE_PRECISION,
      &        MPI_SUM,MPI_COMM_WORLD,IERR)
                ELSE
             DO 70 I=ITASK+1,N,NPES
                P(I) = R(I) + ALPHA*P(I)
      70     CONTINUE
                ENDIF
C                               RMAX = MAX OF RESIDUAL (R)
             RMAXI = 0
             DO 80 I=ITASK+1,N,NPES
                RMAXI = MAX(RMAXI,ABS(R(I)))
      80     CONTINUE
             CALL MPI_ALLREDUCE(RMAXI,RMAX,1,MPI_DOUBLE_PRECISION,
      &        MPI_MAX,MPI_COMM_WORLD,IERR)
C                               IF CONVERGED, MERGE PORTIONS OF X
C                               STORED ON DIFFERENT PROCESSORS
             IF (RMAX.LE.1.D-10*R0MAX)THEN
                IF (ITASK.EQ.0) PRINT *, ' Number of iterations = ', ITER
                CALL MPI_ALLREDUCE(X,R,N,MPI_DOUBLE_PRECISION,
      &           MPI_SUM,MPI_COMM_WORLD,IERR)
                X(1:N) = R(1:N)
                RETURN
             ENDIF
      90 CONTINUE
C                               PCG DOES NOT CONVERGE
         IF (ITASK.EQ.0) PRINT 100
     100 FORMAT(' ***** PCG does not converge *****')
         RETURN
         END


- - - - - - - - - PBAND - - - - - - - - - - - - - - - - - -

         SUBROUTINE PBAND(A,N,NLD,NUD,X,B)
         IMPLICIT DOUBLE PRECISION (A-H,O-Z)
C                               DECLARATIONS FOR ARGUMENTS
         DOUBLE PRECISION A(-NUD-NLD:NLD,*),X(N),B(N)
C                               DECLARATIONS FOR LOCAL VARIABLES
         DOUBLE PRECISION B_(N),LJI,COLUMNI(N)
         INCLUDE 'mpif.h'
C
C   SUBROUTINE PBAND SOLVES THE LINEAR SYSTEM A*X=B, WHERE A IS A
C     BAND MATRIX, WITH COLUMNS STORED CYCLICALLY ON THE AVAILABLE
C     PROCESSORS.
```

```
C
C  ARGUMENTS
C
C                ON INPUT                       ON OUTPUT
C                - - - - -                      - - - - -
C
C   A    - THE BAND MATRIX, DIMENSIONED       DESTROYED.
C            A(-NUD-NLD:NLD,(N-1)/NPES+1)
C          IN THE CALLING PROGRAM.  ELEMENT
C          (I,J) OF THE MATRIX IS STORED IN
C            A(I-J,(J-1)/NPES+1)
C          FOR J=ITASK+1+(K-1)*NPES,K=1,2...
C
C   N    - THE SIZE OF MATRIX A.
C
C  NLD   - NUMBER OF NONZERO LOWER DIAGONALS
C          IN A, I.E., NUMBER OF DIAGONALS
C          BELOW THE MAIN DIAGONAL.
C
C  NUD   - NUMBER OF NONZERO UPPER DIAGONALS
C          IN A, I.E., NUMBER OF DIAGONALS
C          ABOVE THE MAIN DIAGONAL.
C
C   X    -                                   AN N-VECTOR CONTAINING
C                                            THE SOLUTION.
C
C   B    - THE RIGHT HAND SIDE N-VECTOR.
C
C
C- - - - - - - - - - - - - - - - - - - - - - - - - - - - - - - - -
C                              NPES = NUMBER OF PROCESSORS
      CALL MPI_COMM_SIZE (MPI_COMM_WORLD,NPES,IERR)
C                              ITASK = MY PROCESSOR NUMBER (0,1,...,NPES-1).
C                              I WILL NEVER TOUCH ANY COLUMNS OF A EXCEPT
C                              MY COLUMNS, ITASK+1+ K*NPES, K=0,1,2,...
      CALL MPI_COMM_RANK (MPI_COMM_WORLD,ITASK,IERR)
C                              COPY B TO B_, SO B WILL NOT BE ALTERED
      B_(1:N) = B(1:N)
C                              BEGIN FORWARD ELIMINATION
      DO 35 I=1,N
C                              JTASK IS PROCESSOR THAT OWNS ACTIVE COLUMN
        JTASK = MOD(I-1,NPES)
        IF (ITASK.EQ.JTASK) THEN
C                              IF JTASK IS ME, SAVE ACTIVE COLUMN IN
C                              VECTOR COLUMNI
          DO 5 J=I,MIN(I+NLD,N)
            IME = (I-1)/NPES+1
            COLUMNI(J) = A(J-I,IME)
    5     CONTINUE
        ENDIF
C                              RECEIVE COLUMNI FROM PROCESSOR JTASK
        CALL MPI_BCAST(COLUMNI(I),MIN(I+NLD,N)-I+1,MPI_DOUBLE_ PRECISION
     &   ,JTASK,MPI_COMM_WORLD,IERR)
C                              SEARCH FROM A(I,I) ON DOWN FOR LARGEST
C                              POTENTIAL PIVOT, A(L,I)
```

```
          BIG = ABS(COLUMNI(I))
          L = I
          DO 10 J=I+1,MIN(I+NLD,N)
             IF (ABS(COLUMNI(J)).GT.BIG) THEN
                BIG = ABS(COLUMNI(J))
                L = J
             ENDIF
    10    CONTINUE
C                             IF LARGEST POTENTIAL PIVOT IS ZERO,
C                             MATRIX IS SINGULAR
          IF (BIG.EQ.0.0) GO TO 50
C                             I0 IS FIRST COLUMN >= I THAT BELONGS TO ME
          L0 = (I-1+NPES-(ITASK+1))/NPES
          I0 = ITASK+1+L0*NPES
C                             SWITCH ROW I WITH ROW L, TO BRING UP
C                             LARGEST PIVOT; BUT ONLY IN MY COLUMNS
          DO 15 K=I0,MIN(I+NUD+NLD,N),NPES
             KME = (K-1)/NPES+1
             TEMP = A(L-K,KME)
             A(L-K,KME) = A(I-K,KME)
             A(I-K,KME) = TEMP
    15    CONTINUE
          TEMP = COLUMNI(L)
          COLUMNI(L) = COLUMNI(I)
          COLUMNI(I) = TEMP
C                             SWITCH B_(I) AND B_(L)
          TEMP = B_(L)
          B_(L) = B_(I)
          B_(I) = TEMP
          DO 25 K=I0,MIN(I+NLD+NUD,N),NPES
             KME = (K-1)/NPES+1
             LJI = A(I-K,KME)/COLUMNI(I)
             IF (LJI.NE.0.0) THEN
C                             SUBTRACT MULTIPLE OF ROW I FROM ROW J;
C                             BUT ONLY IN MY COLUMNS
                DO 20 J=I+1,MIN(I+NLD,N)
                   A(J-K,KME) = A(J-K,KME) - LJI*COLUMNI(J)
    20          CONTINUE
             ENDIF
    25    CONTINUE
          LJI = B_(I)/COLUMNI(I)
          DO 30 J=I+1,MIN(I+NLD,N)
C                             SUBTRACT MULTIPLE OF B_(I) FROM B_(J)
             B_(J) = B_(J) - LJI*COLUMNI(J)
    30    CONTINUE
    35 CONTINUE
C                             SOLVE U*X=B_ USING BACK SUBSTITUTION.
       DO 45 I=N,1,-1
C                             I0 IS FIRST COLUMN >= I+1 THAT BELONGS
C                             TO ME
          L0 = (I+NPES-(ITASK+1))/NPES
          I0 = ITASK+1+L0*NPES
          SUMI = 0.0
          DO 40 J=I0,MIN(I+NLD+NUD,N),NPES
             JME = (J-1)/NPES+1
             SUMI = SUMI + A(I-J,JME)*X(J)
    40    CONTINUE
```

```fortran
      CALL MPI_ALLREDUCE(SUMI,SUM,1,MPI_DOUBLE_PRECISION,
     &    MPI_SUM,MPI_COMM_WORLD,IERR)
C                               JTASK IS PROCESSOR THAT OWNS A(I,I)
      JTASK = MOD(I-1,NPES)
C                               IF JTASK IS ME, CALCULATE X(I)
      IF (ITASK.EQ.JTASK) THEN
         IME = (I-1)/NPES+1
         X(I) = (B_(I)-SUM)/A(0,IME)
      ENDIF
C                               RECEIVE X(I) FROM PROCESSOR JTASK
      CALL MPI_BCAST(X(I),1,MPI_DOUBLE_PRECISION,JTASK,
     &    MPI_COMM_WORLD,IERR)
   45 CONTINUE
      RETURN
   50 IF (ITASK.EQ.0) PRINT 55
   55 FORMAT (' ***** The matrix is singular *****')
      RETURN
      END
```

References

Alidoust, M. and Linder, J. (2013). Phi-state and inverted Fraunhofer pattern in nonaligned Josephson junctions. *Physical Review B* 87 (6): 060503–1,5.

Alidoust, M., Sewell, G., and Linder, J. (2012). Superconducting phase transistor in four-terminal Josephson junctions. *Physical Review B* 85 (14): 144520–1,8.

Black, F. and Scholes, M. (1973). The pricing of options and corporate liabilities. *Journal of Political Economy* 81 (3): 637–654.

Cvetkovic, S., Fernandez, F., Zhao, A. et al. (1994). Comparison of two interactive finite element programs for analysis of optical and microwave waveguides. *Journal of Lightwave Technology* 12 (7): 1112–1120.

Dang, D.M., Nguyen, D., and Sewell, G. (2016). Numerical schemes for pricing Asian options under state-dependent, regime-switching, jump- diffusion models. *Computers and Mathematics with Applications* 71: 443–458.

Duff, I. and Reid, J. (1983). The multifrontal solution of indefinite sparse symmetric linear equations. *ACM Transactions on Mathematical Software* 9: 302–325.

Duff, I. and Reid, J. (1984). The multifrontal solution of unsymmetric sets of linear equations. *SIAM Journal of Scientific and Statistical Computing* 5: 633–641.

Edsberg, L. (2008). *Introduction to Computation and Modeling for Differential Equations*. John Wiley & Sons.

Fitzgerald, R. and Sewell, G. (2000). Solving problems in computational physics using a general-purpose PDE solver. *Computer Physics Communications* 124: 132–138.

Florescu, I., Liu, R., Mariani, M., and Sewell, G. (2013). Numerical schemes for option pricing in regime-switching jump diffusion models. *International Journal of Theoretical and Applied Finance* 16 (8): 1350046–1,25.

Florescu, I., Mariani, M., and Sewell, G. (2014). Numerical solutions to an integro-differential parabolic problem arising in the pricing of financial options in a Levy market. *Quantitative Finance* 14 (8): 1445–1452.

Flugge, S. (1978). *Practical Quantum Mechanics*, Vols. I, II. Springer- Verlag.

IMSL, Inc. (2010). *IMSL Math Library User's Guide, Version 7*. Houston, TX.

Kadomtsev, B. and Petviashvili, V. (1970). On the stability of solitary waves in weakly dispersive media. *Soviet Physics Doklady* 15: 539–541.

Lu, Z., Tian, E., and Grimshaw, R. (2004). Interaction of two lump solitons described by the Kadomtsev-Petviashvili I equation. *Wave Motion* 40: 123–135.

Patterson, J. and Bailey, B. (2007). *Solid-State Physics*. Springer-Verlag.

Perona, P. and Malik, J. (1990). Scale space and edge detection using anisotropic diffusion. *IEEE Transactions on Pattern Analysis and Machine Intelligence* 12: 629–639.

Sewell, G. (1976). An adaptive computer program for the solution of Div(p(x,y)Grad u) = f(x,y,u) on a polygonal region. In: *The Mathematics of Finite Elements and Applications II* (ed. J. R. Whiteman), 543–553. Academic Press.

Sewell, G. (1985). *Analysis of a Finite Element Method: PDE/PROTRAN*. Springer-Verlag.

Sewell, G. (1988). Plotting contour surfaces of a function of three variables. *ACM Transactions on Mathematical Software* 14 (1): 33–41.

Sewell, G. (1989). An interactive waveguide program. *Proceedings of the Fifth Annual Review of Progress in Applied Computational Electromagnetics*, Monterey, CA. (March 1989), 793–805.

Sewell, G. (2010). Solving PDEs in non-rectangular 3D regions using a collocation finite element method. *Advances in Engineering Software* 41 (5), 748–753.

Sewell, G. (2013). Solving the KPI wave equation with a moving adaptive FEM grid. *Bulletin of Computational Applied Mathematics* 1 (1): 51–67.

Sewell, G. (2014). *Computational Methods of Linear Algebra*, 3e. World Scientific Publishing Company.

Sewell, G. (2015). *The Numerical Solution of Ordinary and Partial Differential Equations*, 3e. World Scientific Publishing Company.

Sewell, G. (2018). Derivation of the Black-Scholes equation from basic principles. *MAA College Mathematics Journal* 49 (3): 212–215. doi:10.1080/07468342.2018.1439634.

Sewell, G. and Cvetkovic, S. (1989). WAVEGIDE–An interactive waveguide program. *Advances in Engineering Software* 11 (4): 169–175.

Smith, B., Boyle, J., Garbow, B. et al. (1974). *Matrix Eigensystem Routines–EISPACK Guide*. Springer-Verlag.

Strauss, W. (2008). *Partial Differential Equations, An Introduction*, 2e. John Wiley & Sons.

Topper, J. (2005). *Financial Engineering with Finite Elements*. John Wiley & Sons.

Zhao, A. and Cvetkovic, S. (1994). Full vectorial simulation of multilayer anisotropic waveguides using an accurate and automated finite element program. *Applied Optics* 33 (24): 5650–5656.

Index

Solving Partial Differential Equation Applications with PDE2D, First Edition. Granville Sewell.
© 2018 John Wiley & Sons, Inc. Published 2018 by John Wiley & Sons, Inc.